Analysis of Energy Systems: Management, Planning, and Policy

T0144459

Series on
ENERGY SYSTEMS: FROM DESIGN TO MANAGEMENT

SERIES EDITOR

Vincenzo Bianco
Università di Genova, Italy

RECENT TITLES

Analysis of Energy Systems: Management, Planning and Policy
Vincenzo Bianco

Analysis of Energy Systems: Management, Planning, and Policy

Edited by
Vincenzo Bianco

CRC Press
Taylor & Francis Group
Boca Raton London New York

CRC Press is an imprint of the
Taylor & Francis Group, an **informa** business

CRC Press
Taylor & Francis Group
6000 Broken Sound Parkway NW, Suite 300
Boca Raton, FL 33487-2742

© 2017 by Taylor & Francis Group, LLC
CRC Press is an imprint of Taylor & Francis Group, an Informa business

No claim to original U.S. Government works

Printed on acid-free paper
Version Date: 20161229

International Standard Book Number-13: 978-1-4987-7739-1 (Hardback)

Library of Congress Cataloging-in-Publication Data

Names: Bianco, Vincenzo, editor.
Title: Analysis of energy systems : management, planning and policy / [edited by] Vincenzo Bianco.
Description: Boca Raton, FL : CRC Press, 2017. | Series: Energy systems from design to management | Includes bibliographical references and index.
Identifiers: LCCN 2016050541 | ISBN 9781498777391 (hardback : alk. paper) | ISBN 9781498777469 (ebook : alk. paper)
Subjects: LCSH: Power resources. | Power resources--Management | Energy policy. | Energy industries.
Classification: LCC HD9502.A2 A55 2017 | DDC 333.79068--dc23
LC record available at https://lccn.loc.gov/2016050541

Visit the Taylor & Francis Web site at
http://www.taylorandfrancis.com

and the CRC Press Web site at
http://www.crcpress.com

Contents

Preface

The analysis of energy systems is of paramount importance in modern societies, since it is fundamental to guarantee a sustainable economic development. It combines technical and economic research with a specific focus on quantitative modeling, in order to optimize the modalities of energy demand and supply globally.

Management, planning, and policy are three key aspects in the study of complex energy systems, which have to be considered at the same time, because of the "multidimensional" nature of energy-related interactions.

Through effective energy management, it is possible to reduce consumption, thus enhancing energy efficiency, and to optimize supply and demand in the medium/short term.

Energy planning, instead, is a fundamental step in pursuing winning long-term strategies in terms of energy supply and demand optimization. Sophisticated instruments and methodologies, ranging from energy demand forecast up to the computation of the optimal supply–demand balance or minimization of pollutant emissions, are available to support complex energy planning.

Similarly, energy policy has a relevant role, because it helps reshape the energy sector and improve its effectiveness by designing and implementing appropriate supporting schemes for the diffusion of the best technologies.

The aim of the present book is to bring together a number of selected contributions regarding the analysis of energy systems from several experts of different countries. The latest research results, innovations, and methodologies are reported in the book in order to support discussion, to circulate ideas and knowledge about the analysis of energy systems.

The most recent trends, such as smart grids and transition from fossil fuels to renewables-based energy systems and distributed generation, are discussed in depth, as briefed here:

- Chapter 1 discusses the nexus between energy transition and sustainability by analyzing possible benefits for the society, the economy, and policy framework necessary to sustain this evolution.
- Chapter 2 proposes an analysis on the role of analytical tools in the context of energy and climate planning. A significant range of tools and methodologies are presented, compared, and discussed in detail.
- Chapter 3 analyzes the role of Energy Service Companies (ESCO) in the energy system and the related policies necessary to stimulate this business. An analysis of the Russian context is proposed, including two case studies on Eurasia Drilling Company and TGT Oilfield Services.

- Chapter 4 discusses the competitiveness of distributed trigeneration, namely, the contemporary generation of heat, power, and cool. An overview of the topic is presented by introducing the main technologies and sectors of application. Furthermore, a methodology to evaluate the competitiveness of the different solutions is presented.

- Chapter 5 focuses on the role of smart grids. In particular, the Northwestern European context is analyzed in terms of economic, environmental, and regulatory issues

- Chapter 6 investigates the problem of renewables optimization on "energy only" power markets. The scope is to analyze the "missing money" problem provoked by the tremendous development of RES in EU, with a focus on the Iberian market.

- Chapter 7 proposes an analysis on the optimal scheduling of a microgrid under uncertainty conditions. An introduction and classification of microgrids is presented and then a methodology to perform the optimization, namely, minimizing the generation cost, is presented and discussed.

- Chapter 8 presents a methodology to perform cost–benefit analysis (CBA) for energy policies. In particular, the chapter focuses on the methodological aspects of the CBA, by illustrating its framework and possible application in the field of energy policy.

- Chapter 9 examines the effect of the implementation of energy efficiency policies in the Middle East and North Africa (MENA) by introducing a benchmarking methodology. The methodology allows comparing the different countries and analyzing their successful policy measures.

- Chapter 10 analyzes the European natural gas market by highlighting the most recent trends, which led to complete reshaping of the sector. The regulatory framework, supply and demand balance, market context, and infrastructure development are discussed in depth.

- Chapter 11 offers a deep insight into the Spanish energy policy with particular emphasis on renewables. A snapshot at the EU level is also given, and the differences between the Spanish and EU context are also highlighted.

The present book aims to be a reference for the academic community, students, and professionals with a wider interdisciplinary background. It provides readers with up-to-date knowledge and innovative ideas for the analysis of energy systems.

Contributors

Vincenzo Bianco
Department of Mechanical, Energy,
    Management and Transportation
    Engineering
University of Genoa
Genoa, Italy

Carmelina Cosmi
Institute of Methodologies for
    Environmental Analysis
National Research Council of Italy
Potenza, Italy

Senatro Di Leo
Institute of Methodologies for
    Environmental Analysis
National Research Council of Italy
Potenza, Italy

Oana M. Driha
Department of Applied Economic
    Analysis
University of Alicante
Alicante, Spain

Tareq Emtairah
International Institute for Industrial
    Environmental Economics
Lund University
Lund, Sweden

Gabriella Ferruzzi
Department of Industrial
    Engineering
University of Naples Federico II
Naples, Italy

Nuno Carvalho Figueiredo
Energy for Sustainability
Faculty of Economics
University of Coimbra
Sustainable Energy Systems
MIT-Portugal
INESCC
Coimbra, Portugal

Giorgio Graditi
Energy Technologies Department
Italian National Agency for
    New Technologies, Energy
    and Sustainable Economic
    Development
Portici Research Center
Naples, Italy

Konstantinos C. Kavvadias
European Commission, Joint
    Research Centre (JRC)
Directorate for Energy, Transport
    and Climate
Petten, the Netherlands

Andrey Kovalev
Institute for Statistical Studies and
    Economics of Knowledge
National Research University
    Higher School of Economics
Russian Federation

Julián López-Milla
Department of Applied Economic
    Analysis
University of Alicante
Alicante, Spain

Michele Moretti
Faculty of Business Economics
Center for Environmental Sciences
Hasselt University
Diepenbeek, Belgium

and

Gembloux Agro-Bio Tech
University of Liege
Gembloux, Belgium

Nurzat Myrsalieva
Energy Consultant
United Nations Development
    Programme
Bahrain

Sylvestre Njakou Djomo
Department of Agroecology
Aarhus University
Tjele, Denmark

Filomena Pietrapertosa
Institute of Methodologies for
    Environmental Analysis
National Research Council of Italy
Potenza, Italy

Liliana Proskuryakova
Institute for Statistical Studies and
    Economics of Knowledge
National Research University
    Higher School of Economics
Russian Federation

Monica Salvia
Institute of Methodologies for
    Environmental Analysis
National Research Council of Italy
Potenza, Italy

Martín Sevilla-Jiménez
Department of Applied Economics
University of Alicante
Alicante, Spain

Patrícia Pereira da Silva
INESCC and CeBER
Energy for Sustainability and
    MIT- Portugal
Faculty of Economics, University of
    Coimbra
Coimbra, Portugal

Jacopo Torriti
School of the Built Environment
University of Reading
Reading, United Kingdom

Steven Van Passel
Faculty of Business Economics
Center for Environmental Sciences
Hasselt University
Diepenbeek, Belgium

and

Department of Engineering
    Management
University of Antwerp
Antwerp, Belgium

Nele Witters
Faculty of Business Economics
Center for Environmental Sciences
Hasselt University
Diepenbeek, Belgium

Georgeta Vidican Auktor
Institute of Economics
Friedrich Alexander University
    Erlangen-Nürnberg
Erlangen, Germany

and

German Development Institute
Bonn, Germany

1

Systemic Interventions to Achieve a Long-Term Energy Transition toward Sustainability

Georgeta Vidican Auktor

CONTENTS

1.1 Multi-Dimensions of Energy System Transition

The main message of this edited volume is that energy systems are multidimensional and require simultaneous interventions at management, planning, and policy level. Given the complex nature of energy-related interactions, a one-dimensional approach to the analysis of energy systems is likely to miss opportunities for achieving long-term sustainable development objectives. With a focus on the link between energy and economic development and on the importance of transforming energy systems toward sustainability, this chapter discusses the need for integrating the analysis of energy systems (simplified here as the system for generation, transmission, and consumption of energy) with other economic and social policies and strategies for enabling a long-term energy transition. The imperative for integration emerges from four main reasons.

First, energy and development are intrinsically related. Economic growth is powered by energy production for fueling industrial development and for sustaining technological innovation; access to energy is essential for improving livelihoods and welfare and for enabling access to markets and entrepreneurship. Thus, an integration of energy policy with industrial and social policy is essential to enabling energy system transformation.

Second, in light of concerns with global climate change, clean energy technologies (in particular renewable energy technologies) need to become more prominent in energy production and consumption. Economic growth based on conventional fuels has historically been associated with increasing levels of environmental degradation. While early industrializing countries have managed to gradually reduce their environmental footprint, developing and emerging economies face massive challenges to decouple economic growth from environmental degradation, especially as related to the scale of investments and know-how. However, opportunities also exist; developing countries could leapfrog industrial development based on fossil fuels by expanding their capabilities in renewables, thus paving the way to a sustainable growth pathway. Such a development process would require, however, a change in the way we produce and consume energy, in the industrial process itself, and the infrastructure for energy generation and distribution.

Third, as conventional energy systems are locked-in into existing infrastructure and power structures (Unruh 2000), institutional path dependencies (e.g., subsidization of fossil fuels) need to be disrupted. The role of the state in this process, in correcting market and coordination failures (Lütkenhorst et al. 2014), is critical. By creating incentives for renewable energy deployment and energy efficiency adoption, a market for low-carbon energy can be enabled, opening up opportunities for local value creation (through job creation, knowledge, and research capabilities). A shift to low-carbon technologies can be, of course, costly in the face of other development requirements (especially for developing and emerging economies), but the co-benefits that can be captured in this process could compensate these costs in the medium and long term.

Fourth, the main question, of course, is how to capture such benefits, how to lock-out the energy system (and the development trajectory) from a fossil fuel-based system and achieve a lock-in into a system based on low-carbon energy technologies. The answer, I argue, lies within an integrated policy approach and long-term planning, based on systemic learning and experimentation. Policy integration across sectors leads to cooperation between stakeholders with different interests and fosters consensus with regard to the direction and sequencing of reforms. Integration can also contribute to opening up new markets for renewable energy technologies with applications in different sectors (e.g., water generation, agriculture, or housing). Long-term planning needs to be based on a national vision for sustainable development, which gives direction and purpose for various initiatives. The narrative framing such a vision should be elaborated by an alliance of diverse stakeholders, representing the civil society, business sector, and policy makers. Systematic learning and experimentation is an essential part of this transition process. Such a systemic and deliberate transformation has almost no precedent in economic development history. Moreover, policy solutions implemented in one context might not be suitable to another. The example of the feed-in tariff (FIT) is a case in point; while this policy instrument has been successfully used in Germany, many developing countries (for instance, India) have found that competitive bidding is more appropriate to their market and institutional context (Altenburg and Engelmaier 2013). Thus, experimentation, reflection, and reassessment of initial objective of strategies are essential for effectively implementing such a transition process, when uncertainties prevail.

This chapter aims to shed light of these issues. As some of these aspects will be discussed in more detail in the subsequent chapters, with reference to different parts of the energy sector, in this chapter I emphasize the systemic nature of the transition process, instead of offering an exhaustive discussion on the subject. The chapter is structured as follows. Section 1.2 discusses the complex link between energy and economic development. The role of the state in developing and guiding the national vision toward sustainability is also discussed. Section 1.3 highlights the co-benefits that can be captured by diversifying the energy system to integrate a larger share of low-carbon energy technologies. Last, Section 1.4 has in focus the need for integrated policy interventions given the complex nature of the process of transformation toward sustainability, emphasizing the importance of learning and systematic implementation of policies. In effect, multidisciplinary and cross-sectoral approaches are critical for capturing value creation in terms of job creation, competitiveness, and poverty reduction.

1.2 Complex Link between Energy and Development

Historically, energy sources have been critical for the development of civilization (Cottrell 1955) and for long-term economic growth and development (Smil 2003). The first industrial revolution is the prime example of how the availability and transformation of energy sources have put into motion the engine of economic growth in our contemporary society (Fouquet 2008, Ayres 2009). Inventions such as the internal combustion engine, creating new ways of moving goods and people and accelerating transportation in our modern world, as well as the steam locomotive (powered by coal or wood) became symbols of modernity and progress by the end of the nineteenth century (Moe 2010, Carbonnier and Grinevald 2011). Later, the railway and air travel accelerated the growth process.

Yet, as Moe (2010) emphasizes, it is the symbiosis between energy and industry contributing to structural change and thus to economic progress. Specifically, without new sources of energy growth in new industries would not have been possible, and, at the same time, without technological change and industrial progress incentives for exploiting and developing new sources would have been minimal. Correlations between human development indicators and electricity consumption per capita show the positive effect that the availability and consumption of energy have on socioeconomic development (GEA 2012).

Energy plays a critical role for poverty reduction. Access to reliable and clean energy has been increasingly recognized to be critical for economic development. Universal access to electricity and modern forms of energy for cooking, as well as switching from traditional solid fuels to cleaner liquid fuels and combustion technologies, is important for developing countries to be able to overcome poverty and support economic growth (GEA 2012).

Access to energy is essential for the delivery of key services, such as education, health, and other social services, for consumption of goods, increasing productivity, and for expanding employment opportunities through industrial development. Currently, there are almost one and a half billion people worldwide without access to electricity (IEA 2010). When considering those who have access only to intermittent sources of energy, this number is much larger.

The relationship between energy and poverty can be characterized by a vicious cycle: poor people lacking access to (cleaner and) affordable energy are often trapped in a cycle of deprivation and limited incomes; at the same time, a large share of their income is used for expensive and unhealthy forms of energy (GEA 2012: 164).

For the preceding reasons, improving access to energy is essential for development. Ensuring access to affordable, reliable, sustainable, and modern energy for all has also been designated by the United Nations as one of the Sustainable Development Goals (SDGs) to achieve the 2030 Agenda. In spite of the difficulty of operationalizing this goal in clear and measurable targets (Loewe 2015), it has a high degree of importance as it emphasizes the need to not only improve access to energy but to also rely increasingly on renewable energy and energy efficiency. As such, it places strong emphasis on transitioning to sustainable energy systems for both developed and developing countries.

However, as it has been widely emphasized, this symbiosis between energy and development owes much to politics, which can either constrain or enable the ability of new energy technologies to transform economies and lay the foundation for long-term economic growth (Moe 2010). The role of politics becomes even more evident in the case of transitions toward a clear energy system.

Specifically, disrupting old pathways (i.e., energy systems based on conventional energy sources) requires overcoming various market failures by creating "policy rents" to unlock the potential for renewable energy and energy efficiency. Market failures, especially emerging from coordination failures, externalities, and the public good nature of environmental quality (see Table 1.1), distort the incentives for investing in clean energy technologies, leading to lock-in in conventional energy systems. Breaking out of this

TABLE 1.1

Typology of Market Failures

Imperfect Competition	Asymmetric Information	Coordination Failures	Public Goods	Externalities
Market power resulting from nonatomistic structures and collusive behavior	Superior information of some market actors (mostly on the supply side)	Obtainable benefits are not being reaped due to lack of coordinated action	Goods that are nonexcludable and non-rival in consumption	Deviation between private and social costs and benefits
		Crucial for creating new and disrupting old techno-economic pathways	Most severe in case of climate change mitigation suffering from "free riding"	Pervasive in environmental pollution, waste management, and natural resource use

Source: Lütkenhorst, W. et al., Green industrial policy: Managing transformation under uncertainty, DIE Discussion Paper 28/2014, German Development Institute, Bonn, Germany, 2014.

lock-in, however, can result in political conflicts as vested interests might act to prevent such transition (Lütkenhorst et al. 2014).

To facilitate the transition toward sustainable energy systems and to overcome various market failures, governments have a particularly important and challenging role to play (Altenburg and Pegels 2012). As opposed to other sectors where the market would select the promising technologies given a certain market size, in the case of clean energy technologies, the government has to set the right incentives (by subsidizing future technologies and investing actively in R&D) to enable deployment and to attract investment in this new sector. The government is also expected to identify new and innovative policies/institutional frameworks that would support such "shift" in the energy system and to reduce coordination failures. For example, incentives need to target the systemic nature of the energy sector, such that investments are made not only in roof-top solar photovoltaic installations, but also in smart grids, new mobility concepts, etc. Governments also need to be concerned with harmonizing national and international policy frameworks (Altenburg and Pegels 2012) so that national policy actions can benefit from the international governance regimes that are constantly evolving in this area. As Mazzucato (2013) argues, the government has always played a critical role in sharing major transitions and directing investments in future technologies and systems. The challenge in the case of energy system is that this process is much more purposive, has a timeline, and is faced with a higher level of complexity and synergies.

1.3 Co-Benefits from the Energy Transition

In order to keep global temperatures rise well below 2°C (UNFCCC 2016), urgent and drastic action needs to be taken at different levels of intervention. According to the IEA (2015) on the global level, the energy intensity of gross domestic product (GDP) and the carbon intensity of primary energy need to be reduced by around 60% by 2050 compared to today, and investments in low-carbon technologies need to be accelerated. While progress in this direction is encouraging, it remains insufficient (ibid).

In their 2007 report, Intergovernmental Panel for Climate Change (IPCC) highlighted that investments in low-carbon energy can deliver various co-benefits (IPCC 2007), among which are improved environmental quality and thus positive effects of health, improved system reliability and energy security, reduced fuel poverty and increased access to energy services, positive impacts on employment and creating new business opportunities through multiplier effects, and substantial savings in energy-related investments.

The energy sector accounted for around two-thirds of the global CO_2 emissions in 2012 (IEA 2015). Thus, an increase in the share of renewables in the energy mix could contribute significantly to pollution reduction. Low-carbon technologies and energy efficiency technologies can deliver large emission reductions across sectors, especially in power generation, but also in transport, industry, and construction. In power generation, renewables can deliver the most emission reductions, almost 200 Gigatons of CO_2 (ibid). Energy efficiency technologies play a much more important role in the industrial sector (i.e., manufacturing, transport sector, and construction). Toward this end, advancements in technology (through investments in research and development) can contribute to achieving further cost reductions that would incentivize large-scale deployment.

Renewable energy and energy efficiency investments can also deliver a set of socioeconomic benefits that can be grouped into macroeconomic effects, distributional effects, energy system–related and additional effects with respect to risk reduction, for example (see Figure 1.1).

Job creation is one such effect that is essential for both developed and developing countries. Employment is necessary not only to sustain social and economic development; it also plays a role in building acceptance for investments in renewable energy and energy efficiency infrastructure. The initial investments costs in clear energy technologies are high and some of these costs might be transferred to consumers through higher energy prices. Thus, if individuals/consumers do not see the direct benefits from such investments, in the form of jobs and welfare effects, support for such a transition in the short term would be lower.

With regard to jobs, in 2014 an estimated 7.7 million jobs (excluding hydro) were registered in the renewable energy sector (IRENA 2015).

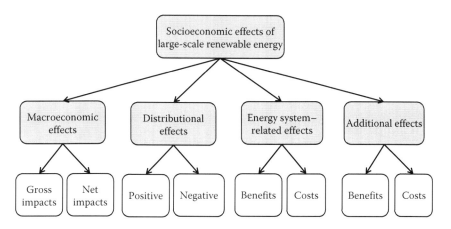

FIGURE 1.1

Types of socioeconomic effects from large-scale deployment of renewable energy. (Adapted from IRENA, The socio-economic benefits of solar and wind energy, EconValue, International Renewable Energy Agency, Abu Dhabi, United Arab Emirates, 2014.)

Not surprisingly, the countries with the largest number of renewable energy jobs were China, Brazil, the United States, India, Germany, Indonesia, Japan, France, Bangladesh, and Colombia. Most jobs were generated by investments in the solar photovoltaics technology (2.5 million), biofuels (1.8 million), wind energy (1 million), and biomass and biogas (ibid). IRENA (2016) finds that doubling the share of renewables would increase direct and indirect employment in the sector to 24.4 million by 2013. Most of these jobs will be generated from fuel supply, installations, and equipment manufacturing.

Overall, doubling the share of renewables in the global energy mix by 2030 would increase global GDP by up to 1.1% or $1.3 trillion. Welfare effects would also increase by 3.7% with regard to economic effects from consumption and investment, social impacts based on expenditure on health and education, environmental effects measured as greenhouse gases (GHG) emissions, and materials consumption.

Private and public investments in clean energy technologies (renewables and energy efficiency) are also likely to create new sources of dynamic competitiveness. As most jobs are likely to be created in installation and manufacturing of parts and components, developing and emerging economies could find opportunities for creating new industries and entering global value chains for renewable energy technologies. Examples such as China, India, and Brazil, but also lower middle-income countries currently deploying renewables at large scale, such as Morocco, illustrate these opportunities.

1.4 Integrated Policy Interventions

To capture benefits from the transition to a sustainable energy system, an integrated policy approach and long-term planning based on systematic learning and experimentation is needed. Integrating the policy process across sectors is necessary to foster synergies and to potentially expand the market for clean energy technologies beyond electricity generation. Specifically, linking energy policy aimed at diversifying the energy mix and improving energy security with industrial policy seeking to achieve economic transition (structural transformation) can contribute to achieving higher levels of decarbonization. Promoting large-scale deployment of renewables and linking these technologies to applications in various sectors, such as housing, agriculture, and transportation, can open up the range of applications and thus enlarge the market. This integrated policy has been referred to as "green industrial policy" defined as "any policy measure aimed at aligning the structure of a country's economy with the needs of sustainable development within established planetary boundaries—both in terms of absorption capacity of ecosystems and the availability of natural resources" (Lütkenhorst et al. 2014: 6). Most policy instruments for promoting energy transitions can be found in the conventional industrial policy toolbox for guiding investments and consumption behavior (see Figure 1.2). These policy tools can be used to promote new pathways, enable innovation in future technologies, support the development of capabilities in the private sector, subsidize clean energy technologies, remove subsidies for fossil fuels, etc. As mentioned earlier, the specifics of green industrial policy tools lie in its cross-sectoral reach, seeking to not only create new markets but to also achieve decarbonization goals within traditional sectors and existing activities.

In addition, to achieve the objective of transitioning to sustainable energy systems and avoid lock-in into high-carbon development pathways, policy interventions need to be based on long-term planning horizons (Lütkenhorst et al. 2014, Hogarth et al. 2015). Long-term planning needs to be supported by a national vision, a broad-based social contract (Lütkenhorst et al. 2014) that gives policy directionality (Mazzucato 2013).

To achieve this broad-based consensus, transformative alliances between stakeholders with different interests are seen to be necessary to overcome potential barriers linked to the political economy of energy transition (Schmitz 2015). In the narrative that frames such alliances (Raven et al. 2016), a strong emphasis on the co-benefits that can be captured by such a transition is necessary to increase the appeal for an alignment of interest. For instance, in Germany, the energy transition has been driven by a coalition of very diverse stakeholders: civil society advocacy groups with a genuine

Information instruments (influence behavior through awareness)	Economic instruments (influence behavior through price)	Regulatory instruments (influence behavior through legality)

Degree of governmental intervention

Low → High

Research and development	Access to resources	Standards (for processes and
Information centers	Taxes	products)
Statistical services	Levies	Property rights/land rights
Awareness campaigns	Tradeable permits	Legally binding targets
Training/education	Direct spending/payments	Quotas
Transparency initiatives	Lending and guarantees	Licenses
Voluntary performance	Insurance (including for bank	Planning laws
targets	deposits)	Accounting systems
Certification/labeling	Government ownership	(mandatory)
(voluntary)	(public–private partnerships)	Copyright and patent
Accounting systems	Public procurement	protection (intellectual
(voluntary)	User fees/charges	property rights)
	Price support or control	Import/export restrictions
		Enforcement

FIGURE 1.2

Types of policy tools available to promote low-carbon transitions. (Adapted from Whitley, S., At cross purpose: Subsidies and climate compatible investment, Overseas Development Institute, London, U.K., 2013.)

green agenda; business circles anticipating the growth of green markets; employers and trade unions alike in sectors benefiting from new jobs (such as wind and solar energy), or in electronic and chemical industries exporting specialized components to green industries worldwide; and regional government and municipalities seeking to strengthen decentralized power structures (Lütkenhorst et al. 2014).

Another relevant dimension is that of systematic learning and experimentation. The complex nature of the transition toward sustainable energy systems (involving high level of uncertainty and risk, synergies and trade-offs between goals and outcomes, and different but interconnected actors) requires a policy process able to respond to these challenges. More than 40 years ago, Schön (1973) claimed that in response to higher uncertainty modifying institutions is not enough; rather "we must invest and develop new institutions which are 'learning systems,' capable of bringing about their own continuing transformation." Systematic policy learning must have two dimensions: learning from others as well as learning over time, where goals and achievements are regularly reviewed

(Lütkenhorst et al. 2014). As Hallsworth (2012) argues, the implications that such an approach has on policy actors is not trivial, as it requires a shift from policy making based on linear thinking to one based on complex adaptive systems. Jones (2011) further explains how when dealing with complex problems, conventional policy tools and approaches are highly unsuitable. One of the most compelling approaches for integrating learning in the policy process is the "learning spiral" developed by the World Bank (Blinderbacher 2010). At its core is an iterative process based on feedback loops that allow the integration of new knowledge in the decision-making process and adds flexibility to revise earlier goals and objectives. Along such learning spirals several policy-making tools need to be used to enhance learning such as horizon scanning, scenario planning, technology forecasting exercises, value-chain analyses, systems mapping, and growth diagnostics.

Last, policy experimentation is critical for enabling and sustaining such long-term transformation of energy systems. Exploring the case of solar energy deployment in India, Altenburg and Engelmaier (2013) find high levels of policy experimentation and learning. In particular, recognizing that the widely used FIT system used to promote renewable energy deployment might not work in India, the Indian government developed a process called competitive reverse bidding in order to identify the "right" level of tariff for solar photovoltaic installations. This system enabled the government to avoid potential risks of faulty allocation and political capture of incentives (or rents) for solar energy technologies. India's National Solar Mission also followed a phased approach that allowed the government to modify guidelines and policies based on experience gained and lessons learned in earlier phases. Thus, this case shows a high level of policy learning and experimentation to tackle the complexity of dealing with a transition toward more sustainable energy system.

Bringing all these elements together, this chapter aimed to emphasize the need to rethink policy approaches when dealing the transformation of energy systems toward sustainability. In particular, the synergistic dynamic between energy, economy and society more broadly requires a strong emphasis on enhancing co-benefits from investments in energy systems, especially the goal is to transition towards more sustainable ways of producing and consuming energy. Last, given the large-scale investments needed for energy systems (based on either conventional or sustainable energy sources) and the impact they have in terms of locking into certain development pathways, policy experimentation, learning, and reflection play a crucial role. Ultimately, to deal with this high level of complexity, and with the political nature of transforming energy systems, the policy process needs to be driven by alliances between stakeholders with a diverse set of interests.

References

Altenburg, T. and T. Engelmaier (2013). Boosting solar investment with limited subsidies: Rent management and policy learning in India. *Energy Policy*, 59: 866–874.

Altenburg, T. and A. Pegels (2012). Sustainability-oriented innovation systems—Managing the green transformation. *Innovation and Development*, 2(1): 5–22.

Ayres, R.U. (2009). *The Economic Growth Engine: How Energy and Work Drive Material Prosperity*. Edward Elgar, Cheltenham, U.K.

Blinderbacher, R. (2010). *The Black Box of Government Learning: The Learning Spiral—A Concept to Organize Learning in Government*. The World Bank, Washington, DC.

Carbonnier, G. and J. Grinevald (2011). Energy and development. International Development Policy, Vol. 2, Graduate Institute of International and Development Studies, Geneva, Switzerland.

Cottrell, F. (1955). *Energy and Society: The Relation between Energy, Social Change and Economic Development*. McGraw-Hill Book Company, New York.

Fouquet, R. (2008). *Heat, Power and Light: Revolutions in Energy Services*. Edward Elgar, Cheltenham, U.K.

GEA (2012). *Global Energy Assessment – Toward a Sustainable Future*. Cambridge University Press, Cambridge, U.K.

Hallsworth, M. (2012). How complexity economics can improve government: Rethinking policy actors, institutions and structures. In: Kay, J. et al. (eds.) *Complex New World: Translating New Economic Thinking into Public Policy*. Institute for Public Policy Research, London, U.K.

Hogarth, J.R., Haywood, C., and S. Whitley (2015). Low-carbon development in sub-Saharan Africa: 20 cross-sector transitions. Overseas Development Institute, London, U.K.

IEA (2010). *World Energy Outlook 2010*. International Energy Agency (IEA), Paris, France. Available at: http://www.worldenergyoutlook.org/media/weo2010.pdf.

IEA (2015). *World Energy Outlook 2015*. International Energy Agency (IEA), Paris, France. Available at: http://www.worldenergyoutlook.org/weo2015/.

IIASA (2014). Energy poverty and development. In: *Global Energy Assessment*. International Institute for Applied Systems Analysis (IIASA), Chapter 2. Available at: http://www.iiasa.ac.at/web/home/research/Flagship-Projects/Global-Energy-Assessment/Chapters_Home.en.html.

IPCC (2007). Fourth Assessment Report: Climate Change (AR4). Intergovernmental Panel for Climate Change (IPCC). Available at: http://www.ipcc.ch/publications_and_data/ar4/syr/en/contents.html.

IRENA (2014). The socio-economic benefits of solar and wind energy. EconValue. International Renewable Energy Agency, Abu Dhabi, United Arab Emirates.

IRENA (2015). Renewable energy and jobs: Annual review 2015. International Renewable Energy Agency, Abu Dhabi, United Arab Emirates.

IRENA (2016). Renewable energy benefits: Measuring the economics. International Renewable Energy Agency (IRENA), Abu Dhabi, United Arab Emirates.

Jones, H. (2011). Taking responsibility for complexity: How implementation can achieve results in the face of complex problems. Working Paper 330. Overseas Development Institute, London, U.K.

Loewe, M. (2015). Goal 7: Ensure access to affordable, reliable, sustainable and modern energy for all. In: Loewe, M. and Rippin, N. (eds.) Translating an ambitious vision into global transformation: The 2030 agenda for sustainable development. DIE Discussion Paper 7/2015, pp. 47–51. German Development Institute, Bonn, Germany.

Lütkenhorst, W., Altenburg, T., Pegels, A., and G. Vidican (2014). Green industrial policy: Managing transformation under uncertainty. DIE Discussion Paper 28/2014. German Development Institute, Bonn, Germany.

Mazzucato, M. (2013). *The Entrepreneurial State: Debunking Public vs. Private Sector Myths*. Anthem Press, London, U.K.

Moe, E. (2010). Energy, industry and politics: Energy, versted interests, and long-term economic growth and development. *Energy*, 35: 1730–1740.

Raven, R., Kern, F., Verhees, B., and A. Smith (2014). Niche construction and empowerment through socio-political work. A meta-analysis of six low-carbon technology cases. *Environmental Innovation and Societal Transitions*, 18: 164–180.

Schmitz, H. (2015). Green transformation: Is there a fast track? In: Scoones, I. et al. (eds.) *The Politics of Green Transformations*. Routledge.

Schön, D. (1973). *Beyond the Stable State: How Certain Schemes to Improve the Human Condition have Failed*. Yale University Press, New Haven, CT.

Smil, V. (2003). *Energy and the Crossroads*. MIT Press, Cambridge, MA.

UNFCCC (2016). Historic Paris Agreement on climate change 195 nations set path to keep temperature rise well below 2 degrees Celsius. Available at: http://newsroom.unfccc.int/unfccc-newsroom/finale-cop21/ (accessed on May 25, 2016).

Unruh, G. (2000). Understanding carbon lock-in. *Energy Policy*, 28: 817–830.

Whitley, S. (2013). At cross purpose: Subsidies and climate compatible investment. Overseas Development Institute, London, U.K.

2

Energy and Climate Planning: The Role of Analytical Tools and Soft Measures

Carmelina Cosmi, Monica Salvia,
Filomena Pietrapertosa, and Senatro Di Leo

CONTENTS

2.1 Introduction

The transition toward a sustainable energy future, fostered by the recent European directives, requires a huge deployment of renewable and energy efficiency to reduce fossil fuels consumption in both energy production and end use in order to respond to the urgent energy and climate challenges. Moreover, energy and climate planning deal with interconnected systems (energy supply, transport, households, etc.) requiring operational planning tools capable to take into account multiple needs and constraints. In this framework, local and regional authorities have a key role in the achievement of the EU 2020 and 2030 Climate and Energy policy objectives, as they are responsible for the definition and implementation of energy policies as well as of infrastructures and services management. However, despite the large number of decision support tools made available by the scientific community, decision-makers are still reluctant to utilize analytical methodologies to support the policy-making process and their little knowledge of local energy systems represents still a main barrier. Besides that, energy awareness of consumers represents a pivotal issue to be taken into account as behavioral changes can strongly contribute to increase energy efficiency. It is therefore necessary to promote the use of consolidated and widespread methodologies for strategic planning as well as to foster voluntary initiatives and the implementation of soft measures through an active engagement of stakeholders.

This chapter summarizes the main concepts of energy and climate planning and aims at providing an overview on analytical tools and soft measures to support energy planners and local authorities in the decision-making processes in order to achieve a substantial improvement of local energy systems performances in compliance with increasingly binding energy and climate targets.

2.2 Energy and Climate Planning: The Policy Framework

The European Parliament adopted in December 2008 the Climate and Energy Package, defining the so-called "20-20-20" policy for 2020. This package and the related targets set for 2020 represent an important first step toward building a low-carbon economy in Europe. In addition to that, the 10-year strategy Europe 2020, launched in 2010, aims at creating the conditions for a smart, sustainable, and inclusive growth and covers employment, research and development, climate and energy, education, social inclusion, and poverty reduction (European Commission, 2016).

The EU 2020 objectives have been translated into national objectives by the national governments recognizing the important contribution of local and regional authorities, as pointed out by the Committee of the Regions (European Union, 2011).

However, Europe looks beyond 2020 defining tighter environmental policy targets and objectives by 2030 and 2050. The EU countries (COM 15 final, 2014) have agreed on a new 2030 Framework for climate and energy, which includes EU-wide targets and policy objectives for the period between 2020 and 2030. Specifically, a 40% cut in greenhouse gas emissions compared to 1990 levels, at least a 27% share of renewable energy consumption and 27% energy savings compared with the business-as-usual scenario should be achieved. This aims to be an intermediate step toward the more ambitious targets set by the Energy Roadmap 2050 (COM 0885 final, 2011), which aims to reduce the EU greenhouse gas emissions (GHG) to 80%–95% below 1990 levels by 2050, increasing innovation and investing in clean technologies and low- or zero-carbon energy. In fact, the transition toward a low-carbon society requires multifaceted interventions aimed to promote energy savings, renewable energy sources, and an efficient use of fossil fuels. These interventions affect different sectors and local/regional competences, from energy supply (e.g., heat and power generation) to the energy demand sectors (e.g., residential and commercial buildings). This framework underlines once more the crucial role played by local and regional authorities to achieve the increasingly tight European targets as they act as "energy consumers and service providers, planners, developers and regulators, advisors, motivators and role models, energy producers and suppliers, buyers" (Energy Cities, 2013). To this issue several initiatives have been promoted to engage local authorities, supporting information sharing and knowledge transfer through networking, such as Climate Alliance (2016), the Covenant of Mayors (2016), Energy Cities (2016), and the ManagEnergy Programme (2016), which involve a large number of EU cities and communities.

In this framework, one of the most significant initiatives is the *Covenant of Mayors* (2016), launched by the European Commission after the adoption, in 2008, of the 2020 EU Climate and Energy Package "to endorse and support the efforts deployed by local authorities in the implementation of sustainable energy policies" (Covenant of Mayors, 2016). In particular, signatories of the Covenant of Mayors commit themselves to achieve at least a 20% reduction of the overall CO_2 emission ("absolute reduction" or "per capita reduction") in 2020 compared to a baseline year set by the local authority. The great success achieved by this initiative has boosted, all around Europe, the development of numerous Sustainable Energy Action Plans (SEAPs) inclusive of energy balances and CO_2 inventories covering key target sectors (buildings and transport, usage of renewable energies, and combined heat and power [CHP]). In particular, as of mid-May 2016, 5985 local authorities signed the Covenant of Mayors and 5253 signatories had already submitted their SEAP (Covenant of Mayors, 2016), bypassing the national level and setting climate goals at the local level where the action plan is implemented (Kjær, 2012).

As an important follow-up, on March 19, 2014, the Covenant of Mayors Initiative on Climate Change Adaptation, *Mayors Adapt,* was set up by the European Commission in order to engage cities in taking action to adapt to climate change (Mayors Adapt, 2016). In particular, the Mayor Adapt initiative aims at increasing support for local activities through the provision of a platform for engagement and networking. Cities signing up to the initiative commit to contributing to the overall aim of the EU Adaptation Strategy by developing a comprehensive local adaptation strategy or by integrating adaptation measures to climate change into existing plans.

This framework led to a new integrated *Covenant of Mayors for Climate & Energy,* launched by the European Commission on October 15, 2015, based on three pillars: mitigation, adaptation and secure, and sustainable and affordable energy (Covenant of Mayors, 2016). Endorsing a shared vision for 2050, signatories commit themselves to reduce CO_2 emissions by at least 40% by 2030 and to adopt an integrated approach to tackling mitigation and adaptation to climate change. The political commitment of signatories is then translated into practical measures and projects through a Sustainable Energy and Climate Action Plan (SECAP), which should be submitted by Covenant signatories within 2 years following the date of the local council decision. Beside the Baseline Emission Inventory, the SECAP will include also a Climate Risks and Vulnerability Assessment whereas the "adaptation strategy can either be part of the SECAP or developed and mainstreamed in a separate planning document" (Covenant of Mayors, 2016).

2.3 Energy Awareness and Behavioral Aspects

Energy consumption is mainly determined by habits, social norms, cultural, and economic factors. In fact, mind-set and culture represent a key aspect to reduce further greenhouse gas emissions and to foster a cultural change toward sustainability. The Low Carbon Economy Roadmap (COM 112 final, 2011) acknowledges that behavioral changes are needed to reach the emissions targets outlining that if behavioral change would occur the targets may be reached at lower costs. Policies and national strategies should therefore be complemented by a set of measures (the so-called "soft measures") aimed at promoting substantial changes in consumer behaviors as pointed out also by the Energy Efficiency Directive (Article 10 "Billing information," Article 12 "Consumer information and empowering programme," and Article 17 "Information and training") (Directive 2012/27/EU, 2012).

An emerging body of literature shows that changes in consumption patterns of households and consumers can achieve considerable reductions in emissions at relatively low costs. Different studies in this field have shown that there is potential for energy savings due to measures targeting behavior (Burchell et al., 2014). The evaluation of the effects of soft measures has shown that feedback on energy consumption can encourage households to save energy, by an average of 5%–15% depending on the measure (Barbu et al., 2013). This is also stated by Heiskanen et al. (2010) who outline the importance to engage energy users in the role of citizens, analyzing "different types of emerging low-carbon communities as a context for individual behavioural change."

A multidisciplinary research aimed at identifying the socio-technical factors that influence residential energy consumption in Belgium (Bartiaux et al., 2006) shows that Belgian households can save, on average, 32% of their energy use for heating and hot sanitary water production and 18.7% of their electricity consumption.

Martiskainen (2007) provides a review of the literature on household energy consuming behaviors and how those behaviors can best be influenced with the goal of reducing energy consumption and carbon dioxide emissions (CO_2). It points out that "majority of energy consuming behaviors are based on habits and routine," "habits need to be broken down and changed by introducing new behaviours," and that to this end "measures such as feedback displays, better billing and micro-generation can help make people more aware of their energy consumption." However, as observed by Hertwich and Katzmayr (2004) changing energy-consuming practices into energy-saving behavior is a very slow process because these practices are inserted in everyday routines. In recent years, there has been growing interest in a range of so-called "soft policy" initiatives such as information campaigns and advisory services that can make more attractive alternative choices that may deeply influence people's aspirations, motivations, and lifestyles.

A fundamental mean to promote energy awareness and improve energy habits is thus to foster citizens engagement, making them aware of their consumption and involving them in energy management practices. Educational centers and public buildings are privileged places to engage people in energy-saving actions in which they can evaluate energy consumption and energy savings deriving from more "conscious" behaviors. This allows promoting co-responsibility and disseminating a sustainability culture that may have a greater and durable impact on the community.

The importance of consumers' behaviors in promoting a sustainability culture and achieving energy and climate targets is widely acknowledged by the EU that in the framework of the former Intelligent Energy—Europe (IEE) Programme supported several projects aimed at turning the concept of "intelligent energy" in practice (IEE Programme, 2016).

The importance of a proactive role of private consumers as a fundamental mean to unlock the consumption patterns and to speed up the transition toward a low-carbon economy is also addressed by several calls on energy efficiency in the current Horizon 2020 EU Programme. This background underlines the importance of utilizing a scientific approach both to support a significant change of collective behaviors toward a smart citizenship and to estimate their impact on energy consumption, environment, and economy.

2.4 Energy Planning Methods and Tools

The transition toward a low-carbon society requires substantial interventions to promote energy savings, renewable energy sources, and an efficient use of fossil fuels. Such a transition needs to address multifaceted issues with huge investments on infrastructures in both energy supply (e.g., H&P generation) and end-use sectors (e.g., residential and commercial buildings refurbishment), requiring multilevel competences for the design, implementation, and coordination of energy and climate plans.

Strategic energy and climate planning is carried out mainly at the regional level, while the implementation of the multilevel goals and policies takes place on a local level through planning, local cooperation, and partnerships (Kjær, 2012). Therefore, local and regional authorities play an ever more crucial role as they are operatively involved in the decision-making processes and are responsible for a large part of the economic structures in their cities and regions.

Energy and climate planning should consider the complexity of energy systems and their interrelated subsystems that encompass all the process chain from primary resources extraction to end-use energy demand through the technology network (Figure 2.1), including also technical, environmental, and socioeconomic constraints (Pfenninger et al., 2014).

Accurate and reliable data on energy production and consumption as well as comprehensive tools are thus necessary for an effective design, implementation, and assessment of energy policies (EC-JRC, 2015). In this framework, data availability still represents a frequent and common concern at local scale (municipal and regional) due to the lack of adequate statistics or databases and the scarce knowledge of existing energy systems. In fact, municipal energy plans and their related energy balances are compulsory only for larger cities (e.g., with more than 50,000 inhabitants, according to Italian Law n. 10/91) and in most of the cases policy-makers cannot rely on

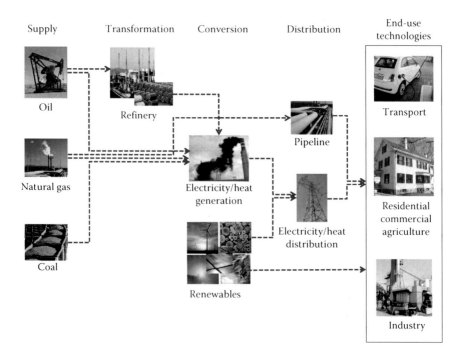

FIGURE 2.1
Example of a reference energy system. (CNR-IMAA elaboration)

a detailed picture of the energy supply and demand within their territories. This lack of data affects the estimation of energy consumption by fuel (electricity, natural gas, diesel, etc.), by demand sector (residential, tertiary, transport, agriculture, and industry), and by end-use (heating, cooling, etc.) with a main concern for the consumption of publicly owned buildings (governmental offices, schools, hospitals, etc.). As pointed out also by ICLEI—Local Governments for Sustainability et al. (2009) "it is fairly easy to gather supply information (how much oil, electricity, gas, etc. the city uses), but it is more difficult to gather information on who uses what energy sources, how they use these and why."

To partially fill this gap, the Covenant of Mayors signatories are required to submit as a first planning document for validation "a typical energy balance in which the energy production and consumption in the considered municipality are estimated in terms of Megawatt per hour (MWh)" (Lombardi et al., 2014) together with a baseline emission inventory (BEI) that "quantifies the amount of CO_2 emitted due to energy consumption in the territory of the local authority (i.e., Covenant Signatory) in the baseline year" (European Union, 2010). The BEI can include also CH_4 and N_2O emissions if specific measures to reduce these greenhouse gases are planned in the SEAP. This prospect of final energy consumptions and greenhouse gas emissions is of fundamental importance in order to describe the baseline situation, identify the key interventions and opportunities to achieve the CO_2 reduction target by 2020 set by the local authorities, and monitor the progress toward this objective. Based on the BEI, the SEAP describes how the Covenant signatory will reach its commitment and defines concrete reduction measures, identifying time frames and assigned responsibilities.

The next sections, mainly based on Salvia et al. (2016), will provide some basic knowledge on energy planning to then focus on the main models and tools that can support the overall planning process.

2.4.1 Concepts of Energy Planning

Energy planning is a cross-sectoral task that involves many activities and a variety of different professional capabilities. There is not a common definition of energy planning although the main meaning is that it aims at "developing long-range policies to help guide the future of a local, national, regional or even the global energy system" (Bathia, 2014) or, in other words, at "ensuring that decisions on energy demand and supply infrastructures involve all stakeholders, consider all possible energy supply and demand options, and are consistent with overall goals for national sustainable development" (IAEA, 2009). Energy planning "is led by the demand for energy services" and is aimed at "optimal energy-efficiency, low- or no-carbon energy supply and accessible, equitable and good energy service provision to users" (ICLEI—Local Governments for Sustainability et al., 2009).

Developing an energy and climate plan is an effective and important first step to reduce greenhouse gas emissions through improved energy efficiency and increased use of renewable energy (Enova, 2008). The ABCD planning process-Awareness; Baseline Analysis; Compelling Vision; Down to Action-(Park et al., 2009) outlines the importance of setting up a planning process in which today's plans and decisions are driven by a vision of successful outcome in the future (backcasting approach) and in which communicating and debating are very important in order to "empower and motivate city employees, citizens, business, and industry" (ICLEI—Local Governments for Sustainability et al., 2009). According to this approach, the planning process consists of four steps that are repeated as the community or organization moves toward sustainability: (A) building awareness and capacity, (B) assessing your baseline, (C) creating your compelling vision, (D) down to action, that is, prioritizing actions to bridge the gap.

There is a general need of guidelines and common practices to support the planning process at municipal scale. According to the Advanced Local Energy Planning (ALEP) methodology (Jank et al., 2005), local energy and climate planning generally starts with a preparatory phase in which the main objectives and boundaries of the planning process are set up as well as the organizational aspects and the key roles (Figure 2.2). In the next step,

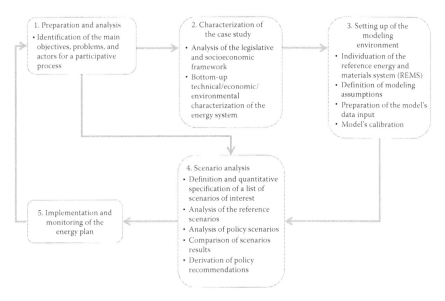

FIGURE 2.2
Overview of the main planning phases, according to the ALEP methodology. (Adapted from Salvia, M. et al., The role of analytical tools in supporting sustainable local and regional energy and climate policies, in: *Proceedings of the International Conference 'Smart Energy Regions'*, Cardiff, U.K., February 11 and 12, 2016, pp. 242–253, http://smart-er.eu/content/proceedings-international-conference-smart-energy-regions-11th-and-12th-february-2016, Accessed May 20, 2016.)

the reference energy system (from energy supply to end-use demands) is in-depth characterized in terms of infrastructures, availability of present and future technologies, energy needs by end-use, and environmental impacts. Then the modeling environment is set up and calibrated on the local case study. A scenario analysis is carried out to analyze alternative pathways of development of the energy system in comparison with a reference scenario (BaU, Business as Usual) in order to devise robust policy strategies. The latest steps of a planning process deal with the implementation of the devised strategies, identifying policy measures and incentives that allow translating the model results into concrete actions, and monitoring the achievement of the planning objectives with possible feedback on the planning strategies, following an iterative approach.

Similarly, the EASY (Energy Actions and Systems for Mediterranean Local Communities) methodology proposes a reference model to define Local Sustainable Energy Strategies with a special focus on Mediterranean cities (Easy IEE Project, 2009). They propose four macro stages, tightly interwoven and complementary, all developed via a cross participation process. First, an assessment stage focused on analyzing the entire energy system in the area and all the related issues, concerns, and weaknesses (Figure 2.3). Second, the planning stage in which the Local Action Plan for Sustainable Energy is developed pointing out strategies, objectives, and priority actions for the local energy system. Third, an implementation stage dealing with two main steps: (1) the development of single projects that put in action the contents of the local action plan and (2) the construction of a beginning scenario (minimal measures, small investments, short timings, small number of local participants) to then arrive, through a series of scenarios, at a final one that is more complex, integrates various projects, has long development times, needs more financing, and requires many local participants. Finally, an evaluation and reporting stage based on the use of a sustainability indicator system in order to monitor the progressive application of the local action plan and evaluate the results obtained so that local administrators can decide whether to adopt corrective actions, review objectives, or restart a new energy planning cycle.

In energy planning, it is important to understand and optimize regional and local energy systems, capturing the dynamics of their interrelated components and assessing the decentralized and variable contributions of

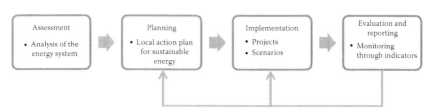

FIGURE 2.3
Conceptual framework for the assessment phase, according to the EASY approach.

renewable energies. The main factors to investigate to analyze the baseline situation are (OECD/IEA, 2013): *Technologies,* in terms of "current status of costs and performance, technology readiness, market penetration and limitations"; *Markets,* dealing with "suppliers, distributors and customers, energy characteristics (production, delivery, storage and consumption) and environmental impacts (air, water and land impacts)"; *Public policies,* as concerns the "current status and requirements of relevant, existing laws and regulations." Therefore, data analysis represents a key aspect in energy modeling, as represented in Figure 2.4.

One of the main challenges is thus to encourage public authorities to adopt, as part of the policy-making process, analytical tools for strategic planning in order to improve steadily the performances of current energy systems and design comprehensive plans with a long-term vision in compliance with the energy and climate targets at European and national scale.

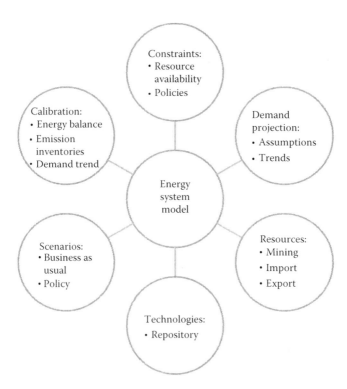

FIGURE 2.4
Main fields of data analysis in energy system modeling. (Adapted from Remme, U., Capacity building through energy modelling and systems analysis, *IEA Experts' Group on R&D Priority-Setting and Evaluation. Developments in Energy Education: Reducing Boundaries,* Copenhagen, Denmark, May 9–10, 2012.)

2.4.2 Energy Models

In the latest decades, energy system analysis has become increasingly important in policy-making (DeCarolis et al., 2012). International bodies and research institutions have developed a wide range of energy models properly designed to help decision-makers in deriving short-term energy and climate strategies within long-term sustainable pathways. In particular, model-based scenario analysis can be very effective in setting up an energy system baseline and to explore "possible future technology deployment pathways" (OECD/IEA, 2013).

Energy system models are typically designed to achieve a balance between accuracy and manageability, their complexity varying from simulation-based spreadsheet models to more elaborate cost optimization models (DeCarolis et al., 2012). Although computer-based models to perform energy system analysis "are being produced at an accelerated pace" (DeCarolis et al., 2012), most of the available models are still unknown to municipalities and local governments that rarely use them to support and assess their policy on energy and climate.

In order to foster the adoption analytical decision support tools for energy and climate planning by public administration, a classification of the widespread available models that highlight their main features can be useful to facilitate the selection of the best suited to the purpose (Van Beeck, 1999).

In the following, a non-exhaustive list of models is presented. Taking into account the many examples of model classification that can be found in the literature (e.g., Allegrini et al., 2015; Connolly et al., 2010; Keirstead et al., 2012; Van Beeck, 1999), the scheme here proposed is based on the following eight key features:

1. *Sectoral coverage*: A first division can be made between **comprehensive models** (that analyze the whole energy system from resources supply to end uses, including all energy transformation processes and taking into account the inter-sectoral relationships) and **sectoral models** (analyzing a single energy sector, e.g., renewable energy generation, district heating/cooling, transport, building energy systems) for a detailed look at the issue.

2. *Geographical coverage*: Depending on the territorial scale of the analysis, the models can be defined as *global* (for world scale analysis) *regional* (for sovra-national level of territorial government analysis, e.g., Europe and North America); *national* (for country analysis), and *local* (for local level analysis, e.g., country regions and more in general provincial and territorial governments) (Van Beeck, 1999).

3. *Time horizon*: That is, the time scale of the analysis: *short term* up to 5 years, *medium term* (3–15 years), and *long term* (50–100 years) (Van Beeck, 1999).

4. *Traditional vs. open source* code models.

5. *Data availability*: Availability of a standard set of data input or a model template.

6. *Routine*: The mathematical background of the modeling approach (e.g., simulation vs. optimization).

7. *Target users*: The level of expertise required to set up and run the model.

8. *Environmental issues*: Modeling of environmental and/or climate issues (e.g., by environmental indicators, emission caps, CO_2 taxes, and external costs).

Comprehensive models are in general long-run energy planning tools, covering a time horizon longer than technologies' lifetime that spans from medium to long term, in order to allow evaluating the effects of technology substitution and the role of emerging technologies.

Sectoral models focus generally on a short/medium time horizon and, in some cases, on a single year (e.g., DIETER, GTMax, and URBS).

Table 2.1 reports a non-exhaustive list of renowned comprehensive models designed for policy assessment at sovra-national and national level that require a high level of expertise in either energy system modeling or scenario analysis. In particular, the MARKAL/TIMES models generator was widely utilized for representing and analyzing energy systems at local scale, demonstrating its effectiveness and scalability (Di Leo et al., 2015; IEA-ETSAP, 2016; Jank et al., 2005; Salvia et al., 2004).

A non-exhaustive list of sectoral models is reported in Table 2.2.

In the latest years, a great attention has been given to *open source tools* (e.g., open models and open data) freely available without a commercial license or a research agreement allowing a greater independence to model developers and users, enhancing quality, transparency, and credibility of models as well as encouraging their utilization for policy advice also among public authorities (local, regional, and national). Moreover, the diffusion of open data cloud computing foster tools and data sharing, reducing the costs, stimulating the research, and creating value-added products build upon their related data streams (MELODIES, 2016).

To this issue, modelers from various universities and research institutes across Europe launched the Open Energy Modelling Initiative (2016a) aimed at boosting the utilization of open source models. In fact, its primary mission is "to enable Open Source energy modelling by providing a platform for collaboration as well as tools along the full value chain of energy economics and energy system models." A Wiki Workspace was also set to collect and update information on the available open data and models (OPENMOD, 2016b).

Several *open models* are currently available both among comprehensive tools (e.g., Calliope, Global Calculator, OSeMOSYS, and TEMOA) and sectoral

TABLE 2.1

Overview of the Main Comprehensive Models

Model Name	Geographical Coverage	Time Horizon	Open Source	Input Database	Routine	Target Users	GHG and Air Emissions
Calliope (Pfenninger and Keirstead, 2015) Free download from http://www.callio.pe/	National and local	Medium to long term	Yes	No	Optimization	Experienced energy system modelers	Yes
EnergyPLAN Free download from www.energyplan.eu	Regional, national, and local	Medium to long term	No	Yes. It provides a set of input data (*Startdata*)	Simulation	Researchers, consultancies, and policy-makers	Yes
Global Calculator Online tool available here: http://uncached-site.globalcalculator.org/	Global	Long term	Yes	Yes	Simulation	Researchers, NGO, policy-makers, and students	Yes
LEAP Free for developing world and students worldwide Commercial for OECD countries: http://www.sei-us.org/leap	Global, regional, national, and local	Medium to long term	No	Yes. It provides a "starter" data set at national level	Simulation	Government agencies, academics, nongovernmental organizations, consulting companies, and energy utilities	Yes
Markal/TIMES Commercial: http://www.iea-etsap.org	Global, regional, national, and local	Medium to long term	No	Yes. It provides a test energy system (Utopia/TIMES DEMO)	Optimization	Experienced energy system modelers	Yes

(Continued)

TABLE 2.1 (*Continued*)

Overview of the Main Comprehensive Models

Model Name	Geographical Coverage	Time Horizon	Open Source	Input Database	Routine	Target Users	GHG and Air Emissions
MESSAGE Commercial: http://www. iiasa.ac.at/web/home/ research/research Programs/Energy/ Download-MESSAGE. en.html	Global and regional	Medium to long term	No	No	Optimization	Experienced energy system modelers	Yes
OSeMOSYS Free download from www. osemosys.org/getting-started.html	Regional, national, and local	Medium to long term	Yes	Yes. It provides a test energy system (UTOPIA)	Optimization	Energy system modelers	Yes
PRIMES It is not sold to third parties, but it is used within consultancy projects undertaken by NTUA (Athens)	Regional and national	Medium to long term	No	No	Simulation	Experienced energy system modelers	Yes
TEMOA http://temoaproject.org/	National	Medium to long term	Yes	Yes. It provides a test energy system (UTOPIA)	Optimization	Energy system modelers	Yes

TABLE 2.2

Overview of the Main Models for Subsystems Analysis

Model Name	Analyzed Sectors	Geographical Coverage	Time Horizon	Open Source	Database	Routine	Target Users	GHG and Air Emissions
Balmorel Free download from http://balmorel.com/	Power and CHP (combined heat and power) sectors	National	Short to long term	Yes	Yes	Optimization	Experienced energy system modelers	Yes
DESSTINEE Free download from http://tinyurl.com/desstinee	Entire energy system with a focus on the electricity system	Regional (Europe)	Medium to long term	Yes	Yes	Simulation	Experienced energy system modelers	Yes
DIETER Free download from www.diw.de/de/diw_01.c.508843.de/forschung_beratung/projekte/projekt_homepages/dieter/dieter.html	Power system	National	Annual, with a long-term perspective	Yes	Yes	Optimization	Experienced energy system modelers	No
EMPS Commercial: https://www.sintef.no/en/software/emps-multi-area-power-market-simulator/	Power systems with a focus on hydro power	Regional (e.g., Nordic countries)	Short to long term	No	Yes	Simulation Optimization	Experienced energy system modelers	Yes

(Continued)

TABLE 2.2 *(Continued)*

Overview of the Main Models for Subsystems Analysis

Model Name	Analyzed Sectors	Geographical Coverage	Time Horizon	Open Source	Database	Routine	Target Users	GHG and Air Emissions
GTMax Commercial: http://www.adica.com/software.html (for American market)	Electric market analysis, generation and transmission investment, regional interconnection and power exchange	National and local	Annual	No	No	Optimization	Electric utilities, transmission companies, regulatory bodies, research institutes, and consulting firms	No
URBS Free download from https://urbs.readthedocs.org/en/latest/	Multicommodity energy systems with a focus on optimal storage sizing and use	Regional, national, and local	Annual	Yes	Yes	Optimization	Research institutes, universities	Yes

models (e.g., Balmorel, DESTINEE, DIETER). In particular, OSeMOSYS and TEMOA adopt the MARKAL/TIMES and MESSAGE paradigm however requiring "a significantly less learning curve and time commitment to build and operate" (Howells et al., 2011).

A common problem in energy modeling is the availability of up-to-date information and the level of data disaggregation to model the reference system as well as the baseline scenario.

To this issue, many energy models contain a kind of "starter data set" that help modelers to build their reference energy model. As an example, LEAP provides the historical energy balance data from IEA-International Energy Agency, emission factors from IPCC, population projections from United Nations, development indicators from the World Bank, and energy resource data from the World Energy Council for over 100 countries (as of November 23, 2011), whereas EnergyPLAN automatically defines a set of input data, called "Startdata" at the beginning of the modeling exercise. Other models such as MARKAL/TIMES, OSEMOSYS, and TEMOA provide an initial simplified reference energy system to be used to build up the model and test the results (e.g., Utopia and TIMES DEMO) (Howells et al., 2011; Hunter et al., 2013).

2.4.3 Tools for Local Administrators

A current operational challenge is to make available user-friendly, open source tools that can be easily transferred to energy planners, policy-makers, and local administrators, after a short "hands-on" training, to support policy design, implementation, and assessment.

Among the many models available to date, Table 2.3 reports a selection of user-friendly decision-making support tools particularly suited for local administrations whose details are described in the following.

The *2050 Calculator* (Department of Energy & Climate Change, 2013) is a user-friendly model developed by the Department of Energy & Climate Change (DECC) of the UK with the aim to explore pathways to reducing greenhouse gas emissions meeting at the same time energy needs and visualizes the effects of behavioral changes on climate change. Its features allow citizens, students, and local administrators to create their own emissions reduction paths. Three versions are currently available to address a broader range of audiences. The 2050 Calculator was designed for the UK energy system, but its approach has also been replicated in other parts of the world with the support of the UK DECC.

The *E2 Tool* (E2 tool, 2015) is a spreadsheet-based tool that can be used to develop energy consumption and GHG emission forecasts for milestone years (2010, 2015, 2020, 2025, and 2030). It is used by local governments in British Columbia (Canada) also to assess the impact of reduction measures at the community scale. The focused sectors are agriculture, industrial buildings, transport (personal vehicle and commercial transportation), residential and commercial buildings, and solid waste. The data input is open source

TABLE 2.3

Overview of the Main Decision Support Tools for Energy and Climate Planning

Model Name	Analyzed Sectors	Geographical Coverage	Time Horizon	Open Source	Database	Routine	Target Users	GHG and Air Emissions
2050 Calculator 1. Web-tool version 2. Simplified My2050 simulation 3. Full Excel version Free download from www.decc.gov.uk/2050	Entire energy system	Global, and national	Medium to long term	No	Yes	Simulation	Researchers, policy-makers, citizens, and students	Yes
E2 tool Can be required to Ramona Mattix—rmattix@rdck.bc.ca or Ron Macdonald—ron.macdonald@stantec.com	Residential and commercial buildings, personal vehicle and commercial transportation, solid waste and agriculture	Local	Medium to long term	No	Yes	Simulation	Policy-makers	Yes

(Continued)

TABLE 2.3 (*Continued*)

Overview of the Main Decision Support Tools for Energy and Climate Planning

Model Name	Analyzed Sectors	Geographical Coverage	Time Horizon	Open Source	Database	Routine	Target Users	GHG and Air Emissions
ICLEI tool Download after the approval of the ICLEI from www.iclei-europe.org/ccp/basic-climate-toolkit	Municipal buildings, vehicle fleet, public lighting, residential, commercial, industry, transport, community waste, and agriculture	Local (municipal)	Annual	No	Yes (average national data and emission factors)	Simulation	Consultancies and policy-makers	Yes
Swiss-Energyscope Free download from www.energyscope.ch/calculateur-energetique/	Entire energy system (Switzerland)	National	Medium to long term	No	Yes	Simulation	Policy-makers, citizens and students	Yes
TRACE Free download from www.esmap.org/node/add/tool-download	Interventions of energy efficiency in transport, buildings, water and wastewater, public lighting, power and heat, and solid waste	Local (municipal)	Short	No	Yes— "Playbook" of tried and tested EE measures	Simulation	Policy-makers	No

to avoid the use of specialized data sets. Key data requirements for building the base scenario include statistics on population and dwellings, energy balances, emissions inventories, and population growth forecasts.

The *ICLEI Europe Basic Climate toolkit* (ICLEI, 2016) allows collecting and systematizing the main energy data and provides GHG emission inventories as final output. These inventories can help local governments to understand the emission paths and to individuate the key priority areas and the achievements of different reduction actions. This tool is utilized by many of the signatories of the Covenant of Mayors to support the elaboration of Sustainable Energy Action Plans. It is made up by Excel spreadsheets that are filled in with two categories of data: local government operations and community inventory. The first category takes into account energy consumption of municipal buildings, vehicle fleet, public lighting, water and sewage, and waste and local energy production, while the other category considers the energy consumption in residential, commercial, industry, transport, community waste, and agriculture sectors. An ICLEI add-in tool was also developed in the frameworks of the South East Europe Project REE RE-SEEties (SEE Programme) (2016) to support local administrations in the calculation of the missing input parameters utilizing proxy variables and information made available by regional or national statistics (Salvia et al., 2014, 2015).

The *Swiss-Energyscope* (Moret et al., 2014) is an online platform developed by the Energy Center of EPFL (Ecole Politecnique Federale de Lausanne) with the aim to support Swiss decision-makers by improving their understanding of the energy system (Gironès et al., 2015). The online platform mainly consists of an energy calculator, enabling users to evaluate the effect of a list of possible choices on the energy future of the country. In particular, it shows the effect of the policy and investment decisions on final energy consumption, total cost, and environmental impact. The modeling approach is currently implemented within an online energy calculator for the case of Switzerland; nevertheless, it can be easily adapted to any large-scale energy system. An online wiki and a MOOC (Massive Open Online Course) allow interested users to acquire a basic knowledge on the energy system and to be guided through the learning process and the use of the calculator itself.

The *Tool for Rapid Assessment of City Energy—TRACE* (TRACE, 2016) is a decision-support system implemented to assist local administrators in identifying opportunities to increase energy efficiency. TRACE was developed by the Energy Sector Management Assistance Program (ESMAP), a global technical assistance program administered by the World Bank, and was designed to involve city decision-makers in the deployment of energy efficiency (EE) measures. TRACE focuses on the municipal sectors with the highest energy use: passenger transport, municipal buildings, water and wastewater, public lighting, power and heat, and solid waste. It targets under-performing sectors, evaluates improvement and cost-saving potential, and helps prioritizing actions for EE interventions. It has been used by 27 cities in Africa, Asia, Europe, Central Asia, and Latin America.

Besides that, plenty of easy-to-use models have been developed to assist developing countries in their planning practices as well as to address different purposes. In particular, the Climate-Smart Planning Platform (2016), developed under the aegis of the World Bank, collects and makes available a set of tools addressed to strengthen decision-making processes on climate-smart planning. It also provides a forum where it is possible to find the tools within an extensive list updated over time and share analyses and modeling experiences.

2.5 Citizens Engagement and Local Action Plans: From Theory to Practice

2.5.1 Citizens Engagement

Energy efficiency is considered a "hidden fuel" that can be highly boosted by citizens' engagement. As a matter of fact, a high level of community engagement encourages behavioral changes that have a positive effect on energy systems performances.

The importance of behavioral changes to reduce end-use energy consumption is also acknowledged by Article 7 of the Energy Efficiency Directive of the European Commission—EED (SWD 451 final, 2013) that consider soft measures eligible for funding like energy and CO_2 taxes.

Soft measures include a broad range of actions among which energy advice, energy audits, energy management, education, training, information campaigns, smart metering, labeling, certification schemes, and capacity building with territorial networks all addressed at increasing citizens' awareness and participation. In fact, they represent a powerful instrument in the communities where all the members identify themselves as an active part of a collaborative environment and joint efforts are made to achieve shared goals.

As their effectiveness is highly dependent on consumers' response and the capacity of triggering long-lasting behavioral changes, it is important to implement a successful participatory process with a clear identification of the objectives, the interested parties (stakeholders), the factors that may hinder the process, the methodology, and the instruments to involve them actively as well as to measure the impact of behavioral measures (Easy IEE Project, 2009).

Stakeholders can deeply influence directly or indirectly the success of policy measures depending on their role and attitude. Therefore, it is necessary to promote a proactive dialogue among the different categories creating a multidisciplinary environment in which everybody can provide their own contribution of creative ideas and solutions through an intensive and concrete engagement.

To this issue it is important to identify a list of possible stakeholders and involve them at the very outset as well as in the crucial phases of any decision-making process.

Among the different techniques, brainstorming and mapping are very useful to focus on stakeholders' macro-categories considering their political, economic, social interests, knowledge and operating experience, benefits and disadvantages, as well as their possible role as supporters and opponents. The stakeholder selection can be completed with a map or a table that can provide a visual overview of the different groups and their role (Dvarioniene et al., 2015).

Stakeholders' function may differ (e.g., preparation of knowledge bases, development and evaluation of ideas). Therefore, the strategies to engage them should be defined taking into account their role, interests, and potential multiplier effect.

An example of stakeholder analysis by macro-categories is reported in Table 2.4.

Among the several available instruments that can foster an active engagement of stakeholders, with particular reference to private citizens, it is worth mentioning two innovative tools to promote citizens' awareness through discussion and experiential learning: the "World Cafés" (Brown and the World Café Community, 2002) and the "Energy Labs" (Dvarioniene et al., 2015).

The **World Café** promotes an innovative learning and exchange of knowledge through informal discussions focused on key issues with a strategic view and has proven particularly useful to promote an active engagement in different contexts (e.g., institutional business, health, education, as well as local communities). The discussions are basically self-managed by the participants within a common framework and under the guidance of some reference questions. The leading idea is to create a work environment that inspires participants and invites them to a free discussion promoting dialogue and cross-pollination paving the way to unconventional changes of people's mind-set (Brown and the World Café Community, 2002).

The **Energy Labs** contextualize the Living Lab approach in the energy framework by promoting citizens' information and engagement to boost the deployment of innovative solutions. The leading idea is to involve operatively stakeholders in a set of activities to develop and elaborate climate protection and energy efficiency concepts by leveraging on ideas, understanding, and practices.

In order to make sure that the Energy Lab contributes to the fulfillment of its overall aim, it is important to get a common understanding of the underlying process. Different types of events can be organized (e.g., expert meetings, workshops, and educational activities). The methodology, the type of event, and the specific purpose should be carefully planned according to the envisaged results. Both top-down and bottom-up methodologies can be used, although a bottom-up approach is recommended to elicit feedback. The participants are usually divided in working groups to support their

TABLE 2.4

Stakeholder Overview by Category

Stakeholder Category	Interests	Involvement Instruments	Envisaged Outcomes
Local, regional, and national authorities public administration	Policy-making Energy policy implementation Energy security Reducing energy bills	Decisional processes Energy management Energy audits Smart meter campaigns Certification schemes Information campaigns Thematic workshops Thematic interviews Consultation events (surveys, World Café, etc.) Dissemination events (e.g., conferences, fairs, energy labs) Web-based tools for information/dissemination/participation	Local action plans Capacity building Optimal resource management
Local, regional institutions (e.g., energy agencies, professional, and experts)	Data and statistics Methodology Networking	Energy management Energy audits Energy consulting services smart meter campaigns Certification schemes Thematic workshops Thematic interviews Consultation events (surveys, World Café, etc.) Dissemination events (e.g., conferences, fairs, energy labs) Web-based tools for information/dissemination/participation	Capacity building Optimal resource management
Academics, research, and learning institutions	Scientific advancements Energy-environmental awareness Networking	Support to decisional processes Smart meter campaigns Certification schemes Information campaigns Thematic workshops Consultation events (surveys, World Café, etc.) Dissemination events (e.g., conferences, fairs, energy labs) Web-based tools for information/dissemination/participation	Capacity building Decisional support methodologies and tools Learning-by-doing capacity building

(Continued)

TABLE 2.4 (Continued)

Stakeholder Overview by Category

Stakeholder Category	Interests	Involvement Instruments	Envisaged Outcomes
Financial Institutions Business, industries, trade organizations (category associations, consortia, chamber of commerce, etc.)	Economic development Networking	Support to overcome the market barriers Thematic interviews Consultation events (surveys, World Café, etc.) Web-based tools for information/participation	New business models Capacity building
Utility suppliers (i.e., electricity, heat, other energy services)	Ensure reliable, affordable, and clean energy supply Networking	Support to overcome the market barriers Smart meter campaigns Certification schemes Energy labeling Thematic workshops Thematic interviews Consultation events (surveys, World Café, etc.) Dissemination events (e.g., conferences, fairs, energy labs) Web-based tools for information/participation	Collective purchasing Optimal resource management Capacity building
NGOs/environmental/consumer associations	Civic engagement Energy-environmental awareness Networking	Smart meter campaigns Information campaigns Thematic workshops Thematic interviews Consultation events (surveys, World Café, etc.) Dissemination events (e.g. conferences, fairs, energy labs) Web-based tools for information/dissemination/participation	Learning-by-doing Smart community Capacity building
Private citizens	Cost-effective energy services Improving quality of life Health and environmental concern	Smart meter campaigns Information campaigns Dissemination events (e.g., conferences, fairs, energy labs) Living labs Consultation events (surveys, World Café, etc.) Web-based tools for interactive participation	Learning-by-doing Increased awareness Prosumers vs. consumers Smart citizenship

active involvement and to provide their findings. It is of utmost importance to get a reporting documentation that will help analyze the outcomes and maximize the contribution of the event.

2.5.2 The Development of a Local Action Plan

A sustainable energy path for a community should address both energy and environmental issues, with concrete policies and measures aimed at ensuring all citizens a secure, widespread, and affordable access to energy services as well as environmental sustainability (ICLEI—Local Governments for Sustainability et al., 2009).

In this framework, local action plans represent strategic documents through which local authorities confirm their commitment to sustainable development contributing effectively to promote energy mix diversification, local and decentralized power supply, renewable use, energy efficiency, and greenhouse gas emissions reduction. To this issue, these plans should be conceived as an evolutionary process with a strategic view in which organizational aspects, decision-making competences, financial instruments, and common sense contribute to achieve measurable targets in a given time frame.

Many examples of model templates for the definition of local action plans can be found in the literature, most of which have been developed in the framework of interregional cooperation projects (e.g., Enova, 2008; ICLEI—Local Governments for Sustainability et al., 2009; RENERGY Project (INTERREG IVC Programme), 2016; RE-SEEties Project (SEE Programme), 2016).

In particular, the Enova's program for Norwegian municipalities, supported by the Intelligent Energy Europe program, provided a guidebook for municipalities that aim to establish their own local energy and climate plan with the aim "to put in place a long-term strategy including an action plan with a clear focus on practical implementation of measures and activities at the local level" (Enova, 2008). This guidebook points out the importance to have plans that include quantitative targets for energy efficiency, heat, and power generation and are based on local renewable energy sources valorization and greenhouse gas emissions reduction taking into account the organizational capabilities of municipalities that represent the most critical factor in moving from planning to implementation.

The available examples of local action plans highlight different planning phases interlinked by several intermediate steps. However, according to Enova (2008) the operational planning steps can be generally ranked into two main phases: (1) strategic planning and (2) implementation and monitoring.

The *strategic planning* defines the backbone of the planning process, namely, the knowledge basis, the local community position, the overall aim, the strategic objectives, and the key interventions. The first step concerns the background analysis for an in-depth characterization of the regional framework and the community aspirations (e.g., policy framework, geographical features, problematic and advanced areas, endogenous resources, energy consumption, EE and RES deployment, greenhouse gas emissions, market uptake, community needs, and involvement). A structured self-assessment should be performed following a standardized approach that allows a repetition and update of the process, the definition of a baseline scenario, and benchmarking. This preliminary analysis is fundamental to identify what is to be achieved, design the overall strategy, and outline the key issues to be addressed. In this phase energy, targets and priorities at different time horizon are also defined as well as energy efficiency and renewable measures that resonate with local/regional community analysis, in order to find out the key interventions that contribute to an advancement in the main dimensions of sustainable development.

The *implementation and monitoring* phase carries out, revises, updates, and disseminates the planning strategy. The main activities concern the definition of the action lines (i.e., the sectors and the measures) and the activities for the achievement of the strategic and operating targets. This phase also includes the communication plan, to underpin the implementation plan and to engage stakeholders at key stages, and the monitoring strategy to verify the effectiveness of the actions and revise the plan accordingly.

Starting from this broad division, a local implementation plan should contextualize the selection of the measures to be implemented in a study area according to its strengths and weaknesses, the local policies in force, the local "enablers," and the impact they may have in the region. The plan should also be a "living thing" with defined responsibilities and programmed review. Taking inspiration from the Model Implementation Plan developed in the RENERGY Project (RENERGY Project (INTERREG IVC Programme), 2016), a reference structure is presented in Figure 2.5.

A "step-by-step guidance" is then provided in Tables 2.5 and 2.6 in which, according to the planning phase, the steps to be taken are listed and briefly described. These tables provide also several examples to focus on the expected outcomes related to each step and to better understand how they can be translated into concrete actions or documents.

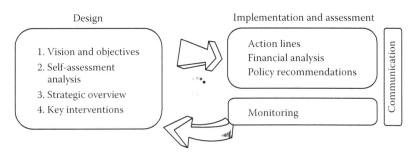

FIGURE 2.5
Reference structure of the planning process, according to the RENERGY methodology.

TABLE 2.5

Main Steps of Strategic Planning

Planning Steps	Objectives	Expected Outcomes	Examples
Vision and objectives	Define overarching objectives and future vision both at community level and in a wider context.	Identification of • Aspirations and role of the community in EE and RES deployment. • Achievable overall objectives at different time scales.	Promote the adhesion of the community to the covenant of mayors. Be renowned as a thriving, low-carbon community by 2020 Boost EE and RES in order to contribute to climate mitigation. Provide a healthy and vibrant environment for citizens and businesses.
Self-assessment analysis	Characterize the study area. Identify strategic objectives and targets for EE and RES increase.	Self-assessment reports SWOT and PEST analysis. Definition of feasible targets according to the initial conditions.	Commit the community to go beyond the 2020 objectives of the COM set by the EU. Reduce carbon emissions by 40% within 2015 and by 40% within 2020.
Strategic overview	Set clear aims and objectives in order to address the challenges faced in achieving the vision.	List and prioritization of aims and corresponding objectives.	*Aim 1:* Demonstrate the relevance of a holistic, integrative, bottom-up process to take local community needs, demands, cultural, and infrastructural characteristics into account. Objective 1: Engage key organizations with measurable carbon saving potential and active community networks to create local community exemplars. *Aim 2:* Emphasize the crucial role of the energy business sector in RES uptake and EE management. Objective 2: Assist local business to reduce energy costs, reduce carbon emissions, and safeguard employment.
Key interventions	Address: • The strategic and territorial challenges. • The vision, aims, and objectives of the community.	List and prioritization of interventions.	Support public and private investment in low-carbon growth Enhance stakeholder participation and cooperation Promote networking in order to promote energy and carbon saving and encourage investment in RES.

TABLE 2.6

Main Steps of Implementation and Monitoring

Planning Steps	Objectives	Expected Outcomes	Examples
Action lines	Translate strategy, aims, and objectives into results.	List and prioritization of strategic lines and corresponding actions.	*SL1 Policy and governance* A1.1: Undertake a study to identify energy demands and the best-suited renewable technologies for energy generation in the local area. A1.2: Create a repository of comparative data. *SL2 Market uptake* A2.1 Support businesses/households to reduce their energy requirements A2.3 Develop public/community/private partnerships to enable a wider uptake of RES and EE. *SL3 Community engagement* A3.1 Engage communities to act as exemplars of bottom-up engagement using existing active networks. A3.2 Engage schools to play a central role in community-led RES schemes.
Financial analysis	Assess the feasibility and cost effectiveness of the proposed interventions.	Identification of the financial constraints. Identification of the potential funding streams and links with ERDF programs. Identification of interventions, which address able to pay, private sector potential, as well as energy poverty challenges.	• Grants to climate protection and energy projects in strategic areas. • Economic regeneration programs.

(Continued)

TABLE 2.6 *(Continued)*

Main Steps of Implementation and Monitoring

Planning Steps	Objectives	Expected Outcomes	Examples
Policy recommendations	Provide a policy framework to accommodate the vision, strategy, aims, and objectives and to support specific actions.	Critical analysis of existing policies to review and/or extend them Reference best practice guide on policy.	• Thematic action plans • Implementation of National Renewable Energy Action Plan (NREAP) on the local level.
Communication	Inform and engage actively the community. Publicize the benefits.	• Identification of the methodology and the key elements for community engagement. • Local action plans owned and implemented by communities.	• Networks for SME. • Information campaigns for householders. • Energy labs.
Monitoring	Verify the effectiveness of actions through benchmarking, political scrutiny, and community reporting. Provide feedbacks for a redefinition of aims and objectives and an improvement of actions.	• Improved understanding of the planning strategy impacts • Review and remedial actions	Development and approval of a climate change strategy.

2.6 Conclusive Remarks

Local and regional authorities are undoubtedly assuming a more and more strategic role in the achievement of national and international energy and climate commitments that call for strategic decision-making. In this framework, it should be underlined that improving the performances of energy systems is a key issue to ensure a future energy and environmental sustainability. In particular, the "development and demonstration of holistic system optimization at local/urban level (Smart Cities and Communities)" represents one of the key themes of the European Commission's strategic energy plan (C 6317 final, 2015).

Decision-makers are thus asked to define policies and roadmaps to face both the energy and environmental challenges and to deploy huge infrastructure investments in a scenario of large future uncertainty (DeCarolis et al., 2012).

This requires a systematic use of analytical tools and procedures in policy design and implementation to provide a benchmarking scenario on which the effectiveness of policies and measures can be assessed and the investments can be carefully planned, according to a backcasting planning approach.

Nevertheless, a multilayer and often not integrated decision-making process, the lack of common protocols, the general complexity of energy models, the lack of data about sectoral consumption, and energy flows across the supply and end-use demand sectors prevent the use of validated methodologies and tools in municipal and regional energy planning. In addition to that, it is important to underline that the key role of stakeholders in responding to the sustainable development challenges cannot be fulfilled by adopting only a top-down policy, unable to seize their needs and aspirations.

A structured community engagement is thus required since the beginning of any process and across all its crucial phases to identify and address social issues and concerns as well as to generate shared innovative solutions. The key issue is, therefore, to foster a transition toward local and regional sustainable energy systems in which "soft measures are an essential lever for the implementation of hard measures" (Energy Cities, 2012) in perfect agreement with the smart cities and communities paradigm. A strategic vision of energy-environmental planning should therefore counterbalance a top-down policy approach with a bottom-up methodology for stakeholder engagement supporting networking and knowledge exchange between all the actors involved in the planning processes. This results in a stronger collaboration between the research community and local authorities that boost the opportunities to participate in inter-regional and transnational cooperation initiatives triggering collective behavioral changes through the sharing of experiences and good practices.

References

Allegrini, J., K. Orehounig, G. Mavromatidis, F. Ruesch, V. Dorer, and R. Evins. 2015. A review of modelling approaches and tools for the simulation of district-scale energy systems. *Renew. Sustain. Energy Rev.* 52:1391–1404.

Barbu, A.D., N. Griffiths, and G. Morton. 2013. Achieving energy efficiency through behaviour change: What does it take? EEA Technical Report No. 5/2013, Publications Office of the European Union.

Bartiaux, F., G. Vekemans, K. Gram-Hanssen, D. Maes, M. Cantaert, B. Spies, and J. Desmedt. 2006. Socio-technical factors influencing Residential Energy Consumption. SEREC CP/52. Part 1: Sustainable production and consumption patterns. Available at: https://www.belspo.be/belspo/organisation/Publ/pub_ostc/CPen/rappCP52_en.pdf [Accessed June 3, 2016].

Bathia, S.C. 2014. *Advanced Renewable Energy System.* Woodhead Publishing India Pvt Ltd, New Delhi, India.

Brown, J. and the World Café Community. 2002. *A Resource Guide for Hosting Conversations That Matter at The World Café.* Whole Systems Associates. The World Café Community Foundation, Greenbrae, CA.

Burchell, K., R. Rettie, and T. Roberts. 2014. Working together to save energy? Report of the Smart Communities Project. Available at: http://business.kingston.ac.uk./sites/all/themes/kingston_business/charmproject/smartcommunities.pdf [Accessed June 3, 2016].

C 6317 final. 2015. Communication from the Commission. Towards an Integrated Strategic Energy Technology (SET) Plan: Accelerating the European Energy System Transformation. Available at: https://ec.europa.eu/energy/en/publications/towards-integrated-strategic-energy-technology-set-plan-accelerating-european-energy [Accessed January 4, 2017].

Climate Alliance. 2016. Climate Alliance Association. Available at: http://www.climatealliance.org/ [Accessed June 08, 2016].

Climate-Smart Planning Platform. 2016. http://www.climatesmartplanning.org/tools.html, Accessed April 29, 2016.

COM 0885 final. 2011. Communication from the Commission to the European Parliament, the Council, the European Economic and Social Committee and the Committee of the Regions. Energy Roadmap 2050. Available at: http://eur-lex.europa.eu/legal-content/EN/ALL/?uri=CELEX:52011DC0885 [Accessed January 4, 2017].

COM 112 final. 2011. Communication from the Commission to the Council, the European Parliament, the European Economic and Social Committee and the Committee of the Regions. A Roadmap for moving to a competitive low carbon economy in 2050. Available at: http://eur-lex.europa.eu/procedure/EN/200241 [Accessed January 4, 2017].

COM 15 final. 2014. Communication from the commission to the European Parliament, the Council, the European Economic and Social Committee and the Committee of the Regions. A policy framework for climate and energy in the period from 2020 to 2030. Available at: http://eur-lex.europa.eu/legal-content/IT/TXT/?uri=celex:52014DC0015R(01) [Accessed January 4, 2017].

Connolly, D., H. Lund, B.V. Mathiensen, and M. Leahy. 2010. A review of computer tools for analysing the integration of renewable energy into various energy systems. *Appl. Energy* 87(4):1059–1082.

Covenant of Mayors. 2016. Covenant of Mayors for Climate & Energy. Available at: http://www.covenantofmayors.eu [Accessed May 13, 2016].

DeCarolis, J.F., K. Hunter, and S. Sreepathi. 2012. The case for repeatable analysis with energy economy optimization models. *Energy Econ.* 34:1845–1853.

Department of Energy & Climate Change. 2013. Exploring how the UK can meet the 2050 emission reduction target using the web-based 2050 Calculator. Available at: https://www.gov.uk/guidance/2050-pathways-analysis [Accessed May 18, 2016].

Di Leo, S., F. Pietrapertosa, S. Loperte, M. Salvia, and C. Cosmi. 2015. Energy systems modelling to support key strategic decisions in energy and climate change at regional scale. *Renew. Sustain. Energy Rev.* 42:394–414.

Directive 2012/27/EU. 2012. Directive 2012/27/EU of the European Parliament and of the Council of October 25, 2012 on energy efficiency. Available at: http://eur-lex.europa.eu/legal-content/IT/TXT/?uri=CELEX%3A32012L0027 [Accessed January 4, 2017].

Dvarioniene, J., I. Gurauskiene, G. Gecevicius, D.R. Trummer, C. Selada, I. Marques, and C. Cosmi. 2015. Stakeholders involvement for energy conscious communities: The Energy Labs experience in 10 European communities. *Renew. Energy* 75:512–518.

E2 tool. 2015. Energy and emissions planning made simple in the Central Kootenays. Available at: http://www2.gov.bc.ca/assets/gov/environment/climate-change/stakeholder-support/ceei/resources/ghgmodelingtool_rdck.pdf [Accessed May 20, 2016].

Easy IEE Project. 2009. Tools and concepts for the Local Energy Planning. Available at: https://ec.europa.eu/energy/intelligent/projects/sites/iee-projects/files/projects/documents/easy_tools_and_concepts_for_local_energy_planning_en.pdf [Accessed June 3, 2016].

EC-JRC. 2015. Location data for buildings related energy efficiency policies. Available at: http://ec.europa.eu/isa/actions/documents/lbna27411enn.pdf [Accessed June 3, 2016].

Energy Cities. 2012. Energy Cities calls for innovative non-technological solutions to be supported within the Horizon 2020 in the spirit of the Intelligent Energy Europe Programme! Available at: http://www.energy-cities.eu/IMG/pdf/Horizon_2020_Energy_Cities_position_paper_final.pdf [Accessed May 16, 2016].

Energy Cities. 2013. Empowering local and regional authorities to deliver the EU climate and energy objectives. Available at: http://www.energy-cities.eu/IMG/pdf/comm_2030_web.pdf [Accessed May 18, 2016].

Energy Cities. 2016. Energy Cities—The European Association of Local Authorities in Energy Transition. Available at: http://www.energy-cities.eu/ [Accessed June 9, 2016].

Enova, S.F. 2008. Municipal energy and climate planning—A guide to the process. Guidebook part 2. Available at: http://www.managenergy.net/download/Norwegian_Guidebook_for_Municipalities.pdf [Accessed May 12, 2016].

European Commission. 2016. 2020 Climate & energy package. Available at: http://ec.europa.eu/clima/policies/strategies/2020/index_en.htm [Accessed May 13, 2016].

European Union. 2010. How to develop a Sustainable Energy Action Plan. Available at: http://www.covenantofmayors.eu/IMG/pdf/seap_guidelines_en.pdf [Accessed May 13, 2016].

European Union. 2011. The role of local and regional authorities in the Europe 2020 National Reform Programmes. Available at: http://cor.europa.eu/en/documentation/studies/Documents/role-lra-europe-2020-programmes.pdf [Accessed May 16, 2016].

Gironès, C.V., S. Moret, F. Marechal, and D. Favrat. 2015. Strategic energy planning for large-scale energy systems: A modelling framework to aid decision-making. *Energy* 90(1):173–186.

Heiskanen, E., M. Johnson, S. Robinson, E. Vadovics, and M. Saastamoinen. 2010. Low-carbon communities as a context for individual behavioural change. *Energy Policy* 38:7586–7595.

Hertwich, E. and M. Katzmayr. 2004. Examples of sustainable consumption: Review, classification and analysis. Industrial Ecology Programme NTNU, rapport 5, Trondheim, Norway. Available at: http://brage.bibsys.no/xmlui/bitstream/handle/11250/242703/126153_FULLTEXT01.pdf?sequence=1 [Accessed May 16, 2016].

Howells, M., H. Rogner, N. Strachan, C. Heaps, H. Huntington, S. Kypreos, A. Hughes, S. Silveira, J. DeCarolis, and M. Bazillian. 2011. OSeMOSYS: The open source energy modeling system: An introduction to its ethos, structure and development. *Energy Policy* 39(10):5850–5870.

Hunter, K., S. Sreepathi, and J.F. DeCarolis. 2013. Modeling for insight using Tools for Energy Model Optimization and Analysis (TEMOA). *Energy Econ.* 40:339–349.

IAEA. 2009. IAEA tools and methodologies for energy system planning and nuclear energy system assessments. Available at: https://www.iaea.org/sites/default/files/INPROPESS-brochure.pdf [Accessed May 16, 2016].

ICLEI. 2016. Basic Climate Toolkit. Available at: http://www.iclei-europe.org/ccp/basic-climate-toolkit/ [Accessed May 18, 2016].

ICLEI—Local Governments for Sustainability, UN-HABITAT, and UNEP. 2009. *Sustainable Urban Energy Planning. A Handbook for Cities and Towns in Developing Countries.* Available at: http://www.unep.org/urban_environment/PDFs/Sustainable_Energy_Handbook.pdf [Accessed June 3, 2016].

IEA-ETSAP. 2016. TIMES Applications at Local Scale. Available at http://iea-etsap.org/indcx.php/applications/local [Accessed June 8, 2016].

IEE Programme. 2016. Consumer behavior projects. https://ec.europa.eu/energy/intelligent/projects/en/project-keywords/consumer-behavior [Accessed June 8, 2016].

Jank, R., J. Johnsson, S. Rath-Nagel, T. Steidle, B. Ryden, H. Skoldberg, V. Cuomo et al. 2005. *Technical Synthesis Report. Annexes 22 J& 33. Energy Efficient Communities & Advanced Local Energy Planning (ALEP).* Faber Maunsell Ltd. on behalf of the International Energy Agency Energy Conservation in Building and Community Systems Programme.

Keirstead, J., M. Jennings, and A. Sivakumar. 2012. A review of urban energy system models: Approaches, challenges and opportunities. *Renew. Sustain. Energy Rev.* 16(6):3847–3866.

Kjær, T. 2012. Local Energy Action plan and the Multilevel Governance. In Seventeenth Reform Group Meeting, Salzburg, Austria, August 28, 2012. Available at: http://www.polsoz.fu-berlin.de/polwiss/forschung/systeme/ffu/veranstaltungen/termine/archiv/pdfs_salzburg/Kjaer.pdf [Accessed May 16, 2016].

Lombardi, M., R. Rana, P. Pazienza, and C. Tricase. 2014. The European Policy for the sustainability of urban areas and the "Covenant of Mayors" initiative: A case study. In R. Salomone and G. Saija (Eds.). *Pathways to Environmental Sustainability: Methodologies and Experiences.* Springer Science & Business Media, pp. 183–192.

ManagEnergy Programme. 2016. Supporting local and regional sustainable energy actions. Available at http://www.managenergy.net/ [Accessed May 30, 2016].

Martiskainen, M. 2007. Affecting consumer behavior on energy demand. Final report to EdF Energy. Sussex Energy Group. SPRU—Science and Technology Policy Research, University of Sussex, Brighton, East Sussex. Available at: https://www.sussex.ac.uk/webteam/gateway/file.php?name=seg-consumer-behaviour-final-report.pdf&site=264 [Accessed May 16, 2016].

Mayors Adapt. 2016. The Covenant of Mayors Initiative on adaptation to climate change. Available at: http://mayors-adapt.eu [Accessed May 13, 2016].

MELODIES FP7 Project. 2016. MELODIES Exploiting Open Data—A FP7 Project. Available at: http://www.melodiesproject.eu/ [Accessed June 8, 2016].

Moret, S., V.C. Girones, F. Marechal, and D. Favrat. 2014. Swiss-energyscope.ch: A platform to widely spread energy literacy and aid decision-making. *Chem. Eng. Trans.* 39(Special Issue):877–882.

OECD/IEA. 2013. Energy technology roadmaps: Data analysis and modelling. Available at: https://www.iea.org/media/training/presentations/Day_1_Session_23_Technology_Roadmaps_Analysis.pdf [Accessed May 16, 2016].

OPENMOD. 2016a. Open Energy Modelling Initiative. Available at: http://www.openmod-initiative.org/ [Accessed June 8, 2016].

OPENMOD. 2016b. Open Energy Modelling Initiative.. Available at: http://wiki.openmod-initiative.org/wiki/Main_Page [Accessed June 8, 2016].

Park, C., M. Purcell, and J. Purkis. 2009. *Integrated Community Sustainability Planning Guide.* The Natural Step Canada, Canada. Available at: http://www.natural-step.ca/sites/default/files/integrated-community-sustainability-planning.pdf [Accessed June 3, 2016].

Pfenninger, S., A. Hawkes, and J. Keirstead. 2014. Energy systems modeling for twenty-first century energy challenges. *Renew. Sustain. Energy Rev.* 33:74–86.

Pfenninger, S. and J. Keirstead. 2015. Renewables, nuclear, or fossil fuels? Scenarios for Great Britain's power system considering costs, emissions and energy security. *Appl. Energy* 152:83–93.

Remme, U. 2012. Capacity building through energy modelling and systems analysis. In IEA Experts' Group on R&D Priority-Setting and Evaluation. Developments in Energy Education: Reducing Boundaries, Copenhagen, Denmark, May 9–10, 2012.

RENERGY Project (INTERREG IVC Programme). 2016. Model Implementation Plan. Available at: http://www.renergyproject.eu/ [Accessed June 8, 2016].

RE-SEEties Project (SEE Programme). 2016. SEE Methodological Toolkit and Criteria for Assessment. Available at: http://www.re-seeties.eu/ [Accessed June 8, 2016].

Salvia, M., S. Di Leo, C. Nakos, H. Maras, S. Panevski, O. Fulop, S. Papagianni et al. 2015. Creating a sustainable and resource efficient future: A methodological toolkit for municipalities. *Renew. Sustain. Energy Rev.* 50:480–496.

Salvia, M., S. Di Leo, F. Pietrapertosa, and C. Cosmi. 2016. The role of analytical tools in supporting sustainable local and regional energy and climate policies. In Proceedings of the International Conference 'Smart Energy Regions', Cardiff, U.K., February 11 and 12, 2016, pp. 242–253. http://smart-er.eu/content/proceedings-international-conference-smart-energy-regions-11th-and-12th-february-2016 [Accessed May 20, 2016].

Salvia, M., C. Nakos, S. Di Leo, and S. Papagianni. 2014. Supporting cities' efforts towards a highly efficient and sustainable resource efficient future: The RE-SEEties integrated toolkit. In N. Marchettini, C.A. Brebbia, R. Pulselli, and S. Bastianoni (Eds.). *Sustainable City IX—Urban Regeneration and Sustainability*, Vol. 2. Wessex Institute of Technology, Southampton, U.K., pp. 1075–1087.

Salvia, M., F. Pietrapertosa, C. Cosmi, V. Cuomo, and M. Macchiato. 2004. Approaching the Kyoto targets: A case study for Basilicata region (Italy). *Renew. Sustain. Energy Rev.* 8(1):73–90.

SWD 451 final. 2013. Commission Staff Working Document. Guidance note on Directive 2012/27/EU on energy efficiency, amending Directives 2009/125/EC and 2010/30/EC, and repealing Directives 2004/8/EC and 2006/32/EC. *Article 7: Energy efficiency obligation schemes.* Accompanying the document Communition from the Commission to the European Parliament and the Council. Implementing the Energy Efficiency Directive—Commission Guidance.

TRACE. 2016. Tool for Rapid Assessment of City Energy. Available at: http://esmap.org/TRACE [Accessed May 20, 2016].

Van Beeck, N. May 1999. Classification of energy models. FEW 777. Communicated by Dr. Ing W. van Groenendaal. Tilburg University & Eindhoven University of Technology, the Netherlands.

3

Energy Innovation Policy: Fostering Energy Service Companies

Andrey Kovalev and Liliana Proskuryakova

CONTENTS

3.1 Introduction

There is a double interplay of the innovation potential of a company, its business model, and the structure of the market that is rarely caught in researchers' focus. This interrelation is of importance to companies working in various segments of the energy sector. In this chapter, we review multiple activities launched by the Russian authorities to foster innovation in technological sectors of the Russian economy. The activities centered on mergers and acquisitions are a major part of this framework in both public and private organizations.

This chapter shows that mergers and acquisitions per se have little influence on the organization's innovation potential. At the same time, a proper strategy based on the clear-cut competitive advantage and the related specialization is a more productive path to foster innovation in an energy company. Such a specialization may be later followed by a series of mergers and acquisitions. However, if the initial step was ignored or neglected, the subsequent mergers cannot trigger corporate innovation. This thesis is illustrated with three cases (Eurasia Drilling Company or EDC, TGT, and Rosgeologiya) described in the following text.

3.2 Research on Energy Service

Energy service companies (ESCos), including those operating in the energy sector, are known for increasing competitiveness (Hirst and Brown, 1990), productivity (Worrell et al., 2003), and innovation activity of their clients. They are also known for offering green, environment-friendly solutions (Dangelico and Pontrandolfo, 2013). However, smooth advancement of products and service provided by ESCos to the market is hampered by a number of barriers (Hirst and Brown, 1990; Jaffe and Stavins, 1994), including market imperfections and asymmetries, excessive transaction costs and institutional factors, as well as underdeveloped markets (Bertoldi et al., 2006; Painuly et al., 2003; Vine, 2005). Certain barriers are more common to developing countries and include institutional barriers, poor energy pricing policies, high transaction costs (Painuly et al., 2003), limited access to capital, and poor management of ESCos (Akman et al., 2013).

Analysis of ESCos and their business models is in the focus of several recently published studies (e.g., Mahapatra et al., 2013; Pätäri and Sinkkonen, 2014), in particular devoted to increasing energy efficiency of residence buildings (Hannon et al., 2013; Lombardi et al., 2016), residential heating (Suhonen and Okkonen, 2013), and the benefits of ESCos for deployment of renewable-based power and heat generation (Borge-Diez et al., 2015; Bustos et al., 2016). Some studies even go as far as analyzing the role of ESCos in advancing the capacity of local energy systems to address social needs (Hannon and Bolton, 2015). Kindström and Ottosson (2016) identified the requirements and barriers for the successful development of local and regional energy service companies. Based on a survey of a few companies, researchers identified essential business model elements. Among the success factors are early support from top managers or incremental approach to the service portfolio, clear targeting of existing customer base, and internally balancing two business models.

Oilfield service companies are not precisely ESCos in the traditional sense of this term. But there is a certain parallel between them, energy service companies provide energy technology and energy-efficient services to general profile companies whose activities related with energy technology, while oilfield service providers are outsourced a number of necessary exploration and production activities being an inherent part of the petroleum sector value added chain. Petroleum companies widely use oilfield services such as geological exploration, enhanced oil recovery, well intervention, and repair. It is interesting to analyze the activity of oilfield service providers as ESCos offering services to the oilfield operator as the owner of the rights. This view helps examine the role of the market structure in shaping the incentives for the petroleum industry to develop cost-efficient and environment-friendly technology and business processes.

Market structure is rarely considered to be one of the key factors determining the effectiveness of each segment of the energy sector and the scale of externalities created by those segments. Some research in this field is available (Azomahou et al., 2008; Lutzenhiser et al., 2001), but it does not concern the Russian petroleum industry that has a number of important distinguishing features such as the "high level of monopolization in domestic energy markets, lack of competition and prohibitively high entry barriers for any link in the value chain" (Proskuryakova and Filippov, 2015: 2800). Therefore, the Russian energy sector can be an illustrative example of the interlink between the structure of a market and its effectiveness.

Formulating precise quantitative criteria for measuring this interlink is a difficult task because of its complexity and due to lack of transparency in the Russian energy sector. For example, Eurasia Drilling Company Limited, one of the largest Russian service companies, was delisted from the London Stock Exchange (Eurasia Drilling Company Ltd., 2015a), and this limits the publicly available information about the company. At the level of smaller energy service companies, lack of transparency is an even more noticeable problem.

Yet the structure of the Russian market points to the differences of the country-specific trends and those in many developed countries. In countries that adhere to sustainable practices, it has become a rule of thumb that areas where competition without negative externalities is possible should be separated from those areas where the operation of a monopoly is natural and reasonable from the economic point of view. In the Russian energy sector, it has become a standard practice to merge organizations or business entities under various types of state control to achieve economy of scope or economy of scale effect.

The ability to shape the structure that maximizes the wealth and matches restrictions on externalities is probably the key element of a consistent economic policy (Li and Yu, 2016). In energy sector and energy markets, understanding and modifying the market structure have an impact on firms' productivity, which is bound to a specific technology (Dai and Cheng, 2016). In some cases, the general principles of economic theories may not coincide with the technological architecture of the engineering systems. Third Party Access (TPA) to district heating water networks (Soderholm and Warell, 2011) is an example of a dilemma where general economic considerations may contradict a specific technology under consideration, and this complicates the theoretical analysis of this problem. Such dilemmas are aggravated by the fact that cross-case comparisons are not always possible. In the given example different heating utility systems work in different conditions, and isolating only one factor of efficiency is a tricky way to compare TPA and non-TPA systems.

It is presumed that small energy service companies face fundamentally different incentives depending on the structure of the markets and the structure of the adjacent markets (petroleum exploration for the case of oilfield services and production, industrial goods, and residential sector for the performance contractors). The energy sector covers only a part of activities related to energy transportation, conversion, and consumption. Negative environmental externalities resulting from these processes depend on organizations having no competence in energy technology and often attributing low priority to energy efficiency and conservation (Fais et al., 2016). Energy service companies are expected to bring professionalism, efficiency, and rigor to the field where energy efficiency would have been paid much less attention otherwise.

3.3 Role of Oilfield Service Companies in Russia

Since 2015, Russian oil and gas industry has been facing several major challenges. Drop of fossil fuel prices, oversupply and increasing competition at the market, gradual depletion of traditional major Russian oilfields, and economic and financial sanctions.

In 2015, the key problems pointed out by the oil and gas companies CEOs and specialists were limited access to capital, lack of qualified specialists, corruption and legislative inconsistencies, and rising costs of field development. According to the survey conducted by Deloitte (2016), among the response measures that companies introduce, the most prominent are effective management of asset portfolios (100% of respondents), attracting partners (38%), increasing management efficiency (38%), and introduction of technological and other innovations (38%). This survey covered energy companies working in extraction and refinery (54%), services (31%), and pure extraction (15%).

On average, 10.9% of Russian manufacturing and service companies in 2013 and 2014 performed innovation activities and companies involved in fossil fuel extraction have slightly lower values—8.6% and 8.5%, correspondingly (Gokhberg et al., 2016). Of all innovations introduced by fossil fuels extraction companies, the majority were of technological nature.

The response of the Russian petroleum companies was to lower research and development and other investments and to increase production. Indeed, in 2015 Russian petroleum companies produced about 10.7 mln barrels per day in average, a historic record since the times of the Soviet Union.

This growth is troubled by several persistent issues. For instance, monopolies are not suited for flexible development of small oilfields benefit from the outdated licensing in Russian petroleum industry while lagging behind leading international petroleum companies in efficiency benchmarks. Low oil recovery factor has been particularly noticeable in the last years, limited to 20%–27% only in 2015, which is unacceptably low.

There are three main categories of oil service companies operating at the Russian market: international service giants, Russian in-house service providers, and independent Russian service companies. All three differ from energy performance contractors in several important aspects. Oilfield service companies work generally in the same area as their clients, and the division of labor between petroleum companies and oilfield service companies depends on the competition between the independent service contractors and in-house service centers.

Oil and gas industry is the most significant sector of the Russian economy. It accounts for the largest share of the Russian budget and the income of the Russian economy in general. Therefore, it is of interest to investigate the mechanisms, problems, barriers, and opportunity windows within this field. Twenty-five years ago, the Russian oil and gas industry inherited a giant but outdated and inefficient Soviet petroleum industry that required restructuring and modernization, which proved to be a long and painful process. The Russian oilfield services inherited many features from the petroleum industry and its history. Just as the petroleum industry in general, the oilfield service market in Russia is highly monopolized. For instance, in Russia's onshore drilling three service companies dominate the market: EDC (22.3%), SurgutNG (21%), and RN-Burenie (17%).

The landscape of Russian oilfield services depends on the state of the Russian oil and gas business in general. As major oil and gas producing provinces have long entered into their maturity stage, the production has decline. Infill drilling aimed to sustain the production on the depleted fields results in similarly declining production per meter drilled in Russia. The currently popular technology of hydraulic fracturing cannot change this trend.

One could expect a gradual shift of field service activities toward greenfield regions, but the glut in the international oil market and the price gulp of 2014 and 2015 slowed down this trend. Until 2014, oilfield service market analysts expected that after a decade of growth, when the Russian onshore market had risen from 10 to more than 20 million m drilled, the growth would slow down but continue. However, the situation in 2016 is much more uncertain. The oil market glut persists, the price did not return to the level of 2013 (as the Russian service companies were expecting), and the major Russian oil and gas companies have cut down their investment programs. Finally, given the federal budget constraints, there is a persistent risk that tax load on the petroleum industry will increase. These factors affect the expectations of the oilfield service companies.

The characteristics of drilling operations in Russia show that onshore well construction and workover was at the level of 15.5 bln in 2015 and was expected to grow to 27.6 bln by 2020. Drilling volumes in Russia are also expected to grow from 22.6 mln m in 2015 to 30.6 mln m in 2020. The average depth of wells in Russia has increased from 2410 to 3185 m over a decade (2005–2014) (Eurasia Drilling Company Ltd., 2015b). These figures testify to the increasing complexity of oilfield services and the advanced competences required from service companies. The aging Russian fleet of drilling rigs will have to be modernized. Therefore, drilling companies will have to attract considerable investments in the short run.

At the same time, the operational efficiency of the leading Russian oilfield service companies lags behind the world leaders. Even though it is growing and companies adopt the best international practices, there is still significant potential for improvement. It is difficult to compare trends across companies and regions as available information is fragmented and somewhat anecdotal. At the same time, gradual growth of indicators such as meters drilled per day illustrates the increasing productivity of oilfield operations.

At present, the Russian petroleum industry seems to be approaching another bifurcation point. Multiple forecasts indicate that oil production in Russia will decline in forthcoming decades. Given the oil price decrease and constant fluctuations, Russian oil and gas companies with inflated costs and mediocre managerial efficiency will have to reconsider the basic principles of their activity.

Another challenge that Russian oilfield service companies face is competition with international companies. International providers of oilfield services have access to or directly develop the most advanced exploration and production technologies that are extensively tested and fine-tuned in international projects all over the world. These companies are larger than their Russian peers and can suggest the full range of oilfield services so that the operator may deal with a single contractor. Their leadership leaves little chance for Russian services in the high-cost market niches. As a result, international oilfield service giants control two-thirds of the Russian market, mostly large-scale high-value projects, so that their Russian competitors have to content themselves with the rest. It comes as no surprise because the choice of service contractors has a considerable influence on the entire project through a number of factors such as the quality of drilling-time log or mud log. Extensive experience may be an exclusive advantage, and if large investments are at stake, the operator prefers to work with a well-known contractor having a long reference list.

Many Russian service companies choose to consolidate in order to lower management costs and gain access to a larger and more stable share of the market. To achieve this goal, consolidation should be followed by a major fundamental restructuring, a long, painful trial-and-error process. Companies have to learn to be efficient. Moreover, even companies with established client base, technological, and management background do not necessarily benefit from a merger. The story of the Halliburton and Baker Hughes deal interrupted in the spring of 2016 is an illustration of this thesis.

From time to time, lobbyists of the Russian service operators call on the authorities to restrict the international service dominance. Such attempts are counterproductive for several reasons. First of all, the unfavorable economic situation has already made client companies switch to low-cost contractors. Second, some Russian oilfield service providers have a "success story." These stories (see, e.g., Zirax Nefteservice, 2016) demonstrate that viable technological start-ups can survive and develop in a hostile business environment with the limited independent up-to-date research and development (R&D) potential. Just as in many other industries, protectionist regulatory policy will conserve inefficiency and undermine the stimuli to develop new technologies instead of seeking the government's protection. Moreover, Russian policy-makers put forward the policy of import substitution, particularly in the energy sector. Even though it is usually declared that import substitution should only be aimed at a small number of critical technologies, various sectoral and industrial lobbyists exploit this leverage to gain access to subsidies, tax reductions, and other state support. In reality, it is impossible to develop (in some cases from scratch) oilfield services similar to international on a tight schedule. The most realistic option to foster Russian oilfield service businesses is to help them achieve the level at which they could cooperate with international companies. International oilfield service leaders may readily outsource and localize some operations if Russian companies prove themselves able to guarantee quality and cost of their services and products. International cooperation also allows Russian oilfield services to get access to the international market.

So far, however, the share of Russian oilfield services had steadily declined. Now that the low oil price period seems to remain at least in the midterm, some small independent Russian oilfield services started to hover on the brink of bankruptcy. There might be several reasons why the glut hit them so hard. First, it is natural that petroleum companies decreased their investments in exploration and production, and the service market has shrunk. Then, there is a specifically Russian problem of monopolization: in-house service centers, subsidiaries, or departments of large energy companies are directly managed or supported by their parent companies of the three mentioned categories (international service giants, Russian in-house service providers, and independent Russian service companies). The small independent ones have the least bargaining power and they are the first to lose the market.

It is hard to predict the dynamics of Russian oilfield service market because of the fluctuating international oil market and unclear Russian economic situation. However, there are some indicative discrepancies. In the absence of comparable replacements of the depleted old deposits, the only way to sustain the overall production level is to intensify the production on the existing sites preferably using cheap and well-established technologies such as infill drilling in combination with different kinds of flooding (such as CO_2 flooding). According to the Russian Energy Minister, sustaining the total production requires increasing infill drilling by 5%–7% annually, given declining production per meter of a well drilled (Tretiyakov, 2014). Moreover, given the quality and average age of the Russian drilling rigs fleet, even maintaining the level of production requires significant investments.

The available data indicate that the market is shrinking. Before the glut, the volume of the Russian oilfield service market had risen to $20–$25 bln. Russian oilfield service market accounted for RUB 700 bln in 2013–2014

($22 bln). The subsequent devaluation of the Russian currency did not change its ruble volume but sharply cut the volume in U.S. dollars. It became a problem for the clients of international oilfield service providers, while petroleum companies could benefit from the ruble devaluation and only enjoyed 11% in ruble revenues (Deloitte, 2015). In 2015 it shrank by 10%, and in 2016 one can expect a further decline.

In the long run, the international oilfield service market is expected to grow because hydrocarbon motor fuels will be consumed in comparable quantities, but simply structured deposits will be largely depleted. Thus, oilfield services will generally become more and more in demand. In Russia, on the contrary, a reversal of the described market trends seems highly unlikely in the short run and, therefore, oilfield services may become a bottleneck hindering the development of the Russian petroleum industry. On the other hand, it is highly likely that the crisis will cause restructuring of the Russian oilfield service market. Generally, one can expect weakening the positions of small independent oilfield service providers, whose market share will be taken by the in-house service providers. The Russian petroleum giants will try to acquire international oilfield service assets, the takeover of which started approximately a year ago with Rosneft acquiring Weatherford subsidiaries. The number of such takeovers will be rather limited because of exhausted financial resources of the Russian petroleum companies.

The problem of externalities is another substantial difference between energy performance contractors and oilfield service companies. The former generally increase energy efficiency and promote energy saving, while the latter only intensify extraction of hydrocarbons that are later processed and used with certain degree of efficiency. In some cases, extraction may become more energy efficient. For example, a steam for SAGD (steam-assisted gravity drainage) (Banerjee, 2012) may be produced using a heat recovery steam generator of a local combined heat and power source. But generally speaking, extraction of hydrocarbons affects the environment. Thus, while energy service is aimed at developing energy conversion and transport processes that generally benefits the environment, oilfield operations may be performed more effectively, but it does not necessarily make them less harmful. This interference has been acknowledged in research (Reis, 1996).

Case 3.1   Eurasia Drilling Company

Being the largest Russian drilling company (in terms of the meters drilled as a measure of the market share), Eurasia Drilling Company Limited (EDC) can serve as an illustration of the entire Russian oilfield service market. The history of the company goes back to 1995 when the company Lukoil-Burenie (Lukoil-Drilling) was founded as the drilling subsidiary of Lukoil, one of the Russian petroleum majors.

As a business entity, EDC was established in 2002 by several Russian and foreign investors, and 2 years later, this company acquired the drilling subsidiary of Lukoil that had become a steadily operating oil-field service subsidiary with established management structure. The company name was changed to Burovaya Companya Eurasia (the Russian Eurasia Drilling Company). The sale of drilling subsidiaries was logical for Lukoil (in 2004 it represented the entire Lukoil's drilling fleet). The company de-invested from a noncore asset, which could make the market more competitive and cut the costs borne by Lukoil. Then, by concentrating on its core business Lukoil increased the efficiency of its operation. This move fits the modern business concept in which oilfield service and petroleum E&P projects are different areas and require different specializations and strategies. Management of daily engineering activity implies that an E&P project operator has to collect and integrate available information about the project, which is a highly innovative technological area including decision support models based on genetic programing, fuzzy logic and neuro-fuzzy models, multi-scenario intelligent optimization, and evolutionary algorithms (Pacheco and Vellasco, 2009). These processes represent a higher level of abstraction from concrete engineering processes in business models and data aggregation as compared with core E&P processes including specific engineering operations and specific models in reservoir simulation, processing data for seismic imaging systems, etc.

The traditional way of dealing with increasingly more complex technology is through a greater specialization* and deeper competence assisted by distribution of risks at every stage of a decision-making process by means of outsourcing operations to a specialized contractor.

Based on the prior development of Burovaya Companya Eurasia, the new management continued the expansion strategy: in the year following the acquisition, the company's share of the Russian drilling market grew to 17%. Two years after the acquisition, the company had roughly 20% of the drilling market and entered the offshore drilling niche. It should be mentioned that the company's offshore interests lie in the region of the Caspian Sea, a more or less traditional area for Russian petroleum companies. Thus, the offshore development did not include deepwater projects, which would have been overly challenging to the Russian petroleum industry.

In the following years, the company continued its merger and acquisition strategy and integrated other Lukoil's service subsidiaries (LUKOIL Shelf Ltd., LUKOIL Overseas Orient Ltd., two West Siberia subsidiaries of Lukoil holding 163 workover rigs, and other Lukoil services). In 2009, the company had more than one quarter of the Russian drilling market,

---

* The growth of service companies may be explained not by increasing specialization, but by the rise of national oil companies. National oil companies possess petroleum reserves but lack access to modern E&P Technologies and, therefore, have demand for oilfield service free from property rights. Oilfield service companies met this demand.

and the merger and acquisition activity continued. After the 2-year-long slump of 2008 and 2009 caused by the economic crisis, the growth continued: the company's share of the market neared one-third, while the total annual length drilled exceeded 6,000,000 m.

Given its close business relations with Lukoil, it was a challenge for the company to diversify its clients in order to avoid a large bargaining power of Lukoil as the major consumer of oilfield services provided by the company. In 2008, EDC started to work with Rosneft in Vankor field, which was followed by a deal with TNK-BP (in 2010) and GazpromNeft (in 2014). Despite its long-term efforts aimed at the diversification of the client pool, Lukoil still remains EDC's major client. In 2015, Lukoil accounted for 56% of the total length drilled by EDC, and a year earlier its share was close to two-thirds of the total length drilled. In fact, it is only recently that the diversification strategy has yielded noticeable results, as GazpromNeft's share has risen to one-third, and it has become EDC's second largest client. At the same time, Rosneft's share even decreased slightly, so the general trend is still mixed. As a result, a drop in drilling activity of Lukoil in 2015, as compared with 2014, still influenced the total length drilled of EDC, which also dropped by 13% (first half of 2015 to first half of 2014). Therefore, the diversification was a forced move and is far from being achieved.

There are several conclusions concerning the growth strategy of EDC. First, the oilfield service market in Russia has been stagnating recently, and given the present oil price trends it will likely stagnate in the near future or longer. There are also high chances that the oilfield market in Russia will shrink, and this decline will hit small oilfield companies in the first place. EDC's growth strategy will also likely be affected by the harsh market conditions.

At present, the potential for any further extension has been largely exhausted. Unlike many smaller companies, EDC has the capacity to increase its efficiency, including efficiency of engineering operations and management efficiency. One of the problems the company is facing is the age of the rig fleet: EDC accumulated a considerable amount of old rigs as a result of its acquisition strategy. The dip in the distribution at the range of middle-age rigs shows that only a few dozen rigs were added to the fleet during the "Lukoil period" of the company's history, which may be explained by the unfavorable market situation (Figure 3.1).

According to the corporate strategy, EDC has lately been paying considerable attention to the development of its offshore division specializing in shallow water drilling. Shallow water reserves account for about 15% of the world oil production. Their average CAPEX per barrel is equal to the CAPEX of expensive traditional onshore reserves, which presently makes these deposits more attractive than deepwater and especially tight oilfields. The cost of tight oil and deepwater projects is a strong incentive

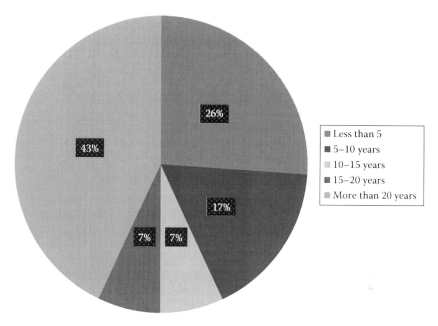

FIGURE 3.1
Age of EDC's rig fleet. (From Eurasia Drilling Company Ltd., EDC land drilling fleet, 2014, available at: http://www.eurasiadrilling.com/operations/rig-fleet/edc-land-drilling-fleet/, last accessed June 7, 2016.)

to intensify the development of technology that could potentially drive the cost down, but Russian companies have not been involved in deep-water projects, and extraction of nonconventional reserves in Russia has been of marginal importance so far. Thus, not only the corresponding technologies are an engineering challenge to Russian developers, but also the demand for such technologies in Russia is marginal. The decision of EDC to focus on shallow water drilling appears logical. On the other hand, maintaining production in old fields gradually becomes more expensive and Russian petroleum companies could benefit from a lower cost-per-barrel ratio that requires cost-efficient drilling technology, such as sidetrack drilling.

This goal may also be a challenge. The situation in Russian drilling is somewhat similar to the middle-income trap known in economics (Agénor and Canuto, 2015). In 2014, the three leading petroleum producers demonstrated the total global drilling volumes shown in Figure 3.2.

While the volumes are almost identical, the trends differ. Both the number of wells drilled and length drilled point to the same fact: the Saudi Arabia producers have access to easy-to-extract resources, for which the necessary amount of oilfield service per extracted barrel is low. The U.S.

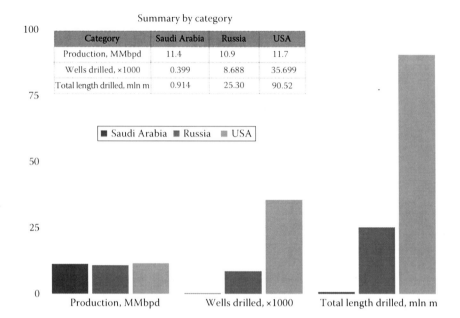

FIGURE 3.2
Drilling volumes that provided the market share of hydrocarbons in 2014 (the number of wells includes sidetracks). (From Kibsgaard, P., *Scotia Howard Weil 2015 Energy Conference*, New Orleans, LA, Schlumberger, March 23, 2015, available at: http://www.slb.com/news/presentations/2015/~/media/Files/news/presentations/2015/Kibsgaard_Scotia_Weil_03232015.ashx, last accessed June 7, 2016).

producers benefit from access to cost-efficient oilfield service operations that assure the same production volumes by means of a larger oilfield service use. The Russian petroleum industry has both disadvantages: it does not have reserves like in Saudi Arabia, and it has no access to equally cost-efficient local oilfield service industry.

Case 3.2   TGT Oilfield Services

Getting into the international oilfield service market is an important measure of success for any oilfield service company. The key is the ability to compete with experienced and cost-efficient international oilfield service companies and local contractors that have a better understanding of their own home market including experience in local supply chain and procurement management to the knowledge of typical local geological formations. A company needs a clear-cut competitive advantage based on managerial experience or technological efficiency giving a competitive edge over the local analogs to come into steady operation abroad.

This logic may not work for Russian technological start-ups. In the generally hostile Russian business environment, where start-ups have to break through multiple financial, administrative, and market barriers, establishing a totally new technological company working for foreign clients may be as difficult as establishing business relationships with residents. At the same time, a company can benefit from the devaluated national currency and lower HR costs.

TGT is an example of such a company. It offers services based on the proprietary logging solutions. The company was founded in 1998, and thus, unlike many other Russian oilfield services and petroleum companies, it did not inherit shabby assets of the Soviet petroleum industry. The company did not have to invest in the renovation or modernization of old equipment and instead can build its organizational structure and technological processes based on best available solutions. Afterward, having established business abroad, such start-ups may shift their focus back to the Russian market and leverage their operational practices based on the acquired understanding of both modern technology and Russian specifics.

The strategy of TGT (predetermined by its specialization) was successful for several reasons: its area did not involve considerable investments in industrial production and depended more on R&D-intensive analysis and computer modeling. This specialization could leverage the high qualification of R&D engineers in Kazan (the hometown) and avoid the problems related to industrial production and operation of heavy E&P machinery, such as drilling rigs. The logging-based business model of the company somewhat resembles a highly mobile IT business: it is not attached to capital-intensive infrastructure (pipelines and refineries) or equipment (rigs), and a significant part of its capitalization is associated with the accumulated engineering background and know-how of the staff. The fact that the company could relocate to United Arab Emirates (UAE) demonstrates this trait.

Having established the core logging service, the company continued R&D activity and presented advanced logging tools, such as high-precision temperature logging, and a number of adjacent services (leak detection, corrosion assessment, etc.); the company started to develop reservoir simulation tools relying on the validation methods based on the well data acquisition tools offered by the company earlier. It is the classical expansion strategy of a company leveraging its core competences to diversify its business.

Having once invested in research, IT infrastructure, and software design, the company does not have to bear those costs fully in the future (except for some maintenance and modernization expenditures): with fully functional software and trained personnel, the marginal cost of working out an additional hydrodynamic model is close to zero. This does not entirely protect the company from unpredictable fluctuations of the oilfield service market inflicted by the unstable trends in the oil business, but it creates a much safer business environment than the one faced by many drilling contractors of similar size in present Russia.

Despite seeming straightforwardness, mathematical modeling underlying numerical simulations of petroleum reservoirs is a complicated process that requires both hard and soft skills. The R&D team ought to have sufficient mathematical and computer qualifications necessary to work with mathematical models, but the set of equations itself is by no means special. It is the specific relevant properties of a reservoir (physical properties, initial and boundary conditions, etc.) that make a model unique. The quality of a computer model, automatic control system, or database resulting from the mathematical model depend a lot on the skills and competences in retaining required phenomena (formation damage, fissuring, wetting for the chemicals used, etc.) while still complying with limited computational complexity and reasonable validations procedure. Such R&D skills, once acquired, create an entry barrier protecting the developer from potential competition. Although the quality of Russian education in natural sciences has declined over the past decades, the history of TGT shows that there are still educated and experienced professionals that are able to support the development of a newly established engineering start-up.

Case 3.3   Rosgeologiya

The third case analyses a company at the geological survey market. This area was a priority in the Soviet times. The dissolution of the Soviet Union and the subsequent decline of the Soviet economy had a negative impact on geological activity. This impact has not been fully bridged so far. There already are petroleum companies that conduct multiple exploration projects, but most experts agree that geological exploration should be intensified in Russia. Insufficiency in this respect may entail considerable risks to the Russian gross petroleum production. Basin geological modeling (Wangen, 2010), as exemplified with the activity of the corporation Rosgeologiya, is an indication of this insufficiency.

Rosgeologiya is a diversified Russian holding that provides geological services. According to the declared priorities and mission statement, Rosgeologiya is similar to the U.S. Geological Survey (USGS). Both organizations were founded in the late nineteenth century: the history of USGS started in 1879, and the history of Rosgeologiya may be traced back through multiple reorganizations to the Geological Committee established in 1882 in the Russian Empire. However, there is a difference between these agencies: USGS is a scientific organization. Geological research is impossible without extensive acquisition of basic geological field data, and USGS activity implies outsourcing to external contractors activities such as observation of well drilling or cooperation with an extensive network of national and international organizations of 564 partners in 2016.

Rosgeologiya, on the contrary, itself conducts 75% of stratigraphic drilling (called parametric drilling in Russia), which is covered by the JSC Nedra being a part of Rosgeologiya. Forty years ago, it was Nedra that drilled the 12,345 m long Kola Superdeep borehole. Kola Superdeep became the deepest well, surpassing Bertha Rogers* drilled in Washita (Oklahoma) by Lone Star Producing Co. while exploring oil and gas (Oklahoma Corporation Commission, 2014). Another representative project of Nedra is Jen-Yakhinskaya superdeep, a sedimentary wellbore (8250 m). At present, this wellbore is rivaled by Rosneft "Sakhalin-1" project with its 13,500 m deep well with horizontal part 12,033 m at Chaivo oilfield (Rosneft, 2015). Such projects demonstrate that present-day oilfield operators can work under demanding geological conditions and cope decently well with this level of complexity. In other words, the most complicated exploration or stratigraphic wells should not be managed as a new Manhattan project. State support may be helpful, but establishing another inflated and sluggish state corporation is not necessary.

USGS acts as a scientific organization that acquires and collects information by exchanging and purchasing data as well as operational activities via an extensive cooperation network. Its Russian analog works in a different way. Russian authorities chose to merge and centralize geological exploration organizations and keep them under state control. In 2011, the state-owned corporation Rosgeologiya gained control over 38 specialized public enterprises. The resulting holding provides a wide range of geological services including mapping and geodetic surveying, geophysical investigations, marine geology, parametric drilling, and more.

Unlike parametric drilling, basin geological modeling is an area that may require an active participation of a systemic moderator.

Basin modeling, as a mathematical description of the geological evolution of a sedimentary basin over a geological period, can be traced back to the late 1970s. Since then the research in this field has been driven by the petroleum industry that considered basin modeling to be a promising tool for exploration. As a result, basin modeling has developed into a sophisticated method. It absorbed geology, geomechanics (and geophysics in general), and chemistry. No wonder modern basin modeling relies on computer simulation, which adds numerical mathematics to the pool of necessary disciplines but allows to reduce investment risks in petroleum E&P projects. Thus, the use of such software can measure the leadership (or the lag) of national geological services.

There are a number of software products designed and continuously developed, such as Beicip Franlab TEMIS (Beicip Franlab, 2016) and PetroMod Petroleum Systems Modeling by Schlumberger (2016). Despite a declared priority of import substitution and self-sufficiency,

---

* Bertha Rogers went through sedimentary formations, while Kola Superdeep cut through the Baltic shield.

Russian developers have not demonstrated similar level products, although up-to-date basin modeling is an important element of replenishing Russian petroleum reserves. There are only a few groups working in this field (Ismail-Zadeh et al., 2016; Malysheva, 2015).

The tools available in Rosgeologiya seem outdated. For example, the corporation announced in April 2016 that one of its subsidiaries (a research institute) will perform basin 1D and 2D modeling for the region of Udmurtiya (Rosgeologiya, 2016). The use of 1D and 2D models is indicative of the development rate. Hantschel and Kauerauf asserted that even though "most models under study were first performed in 2D <...> they were rarely used in practical exploration studies as horizontal petroleum migration in the third dimension cannot be neglected" (Hantschel and Kauerauf, 2009: 16). As a result, a new generation of programs released in the late 1990s included 3D modeling functionality, which has been widely used since then. Simpler models still can be used for describing structurally simple basins. The current world trend is the opposite: the emphasis is shifted toward modeling increasingly more complex structures in 3D and 4D as well as the integration of the corresponding software with other petroleum.

Petroleum exploration in Russia is haunted by a number of technological problems that need to be resolved, but petroleum exploration activities in Russia are under crossfire. On the one hand, most Russian petroleum companies cut costs that do not generate short-term cash flows. On the other hand, the government cannot afford a larger financial support either.

For Russian authors, it is traditional to oppose company-based exploration activities with exploration activities organized and funded by state agencies. But the activity of the U.S. Geological Survey demonstrates that the government agency can outsource survey projects to private companies. There appears to be a more productive approach though. Just as projects in petroleum exploration, design and development of new materials is both complex and systemic activity of strategic importance which requires government intervention. The intervention may intend to intensify R&D and the commercialization of its result. The Materials Genome Program is an emblematic example of an inter-organizational collaborative strategic initiative intended to foster the development of new materials (U.S. National Science and Technology Council, 2011). In this case, the open innovation approach is promoted as the basic framework for achieving a synergetic result.

Merging former Soviet scientific institutions and geological exploration organizations under the state control does not produce a similar change. Inability to implement this or the alternative model productively combining the strengths of corporate and state-controlled exploration activities is an indicative symptom of the inefficient organization of R&D.

3.4 Discussion

The three cases described in this chapter (EDC, TGT, and Rosgeologiya) demonstrate three possible strategies of Russian energy companies:

1. Merger and acquisitions (Rosgeologiya).
2. Specialization followed by expansion (TGT).
3. A combination of (1) and (2): EDC as LUKOIL spin-off was an example of specialization followed by a series of acquisitions.

Of the three mentioned cases, the TGT case seems the most successful from the point of view of innovation potential. The company developed a number of high-tech products and services that are currently offered to Russian and other customers abroad. These products and services are based on the proprietary technology that is undoubtedly innovative.

Merger and acquisition is a popular strategy in Russia, and the energy sector is not an exception. It has been already shown that it may not be successful in some cases (Kovalev and Proskuryakova, 2014). The case of Rosgeologiya testifies to the same thesis. Merging organizations lagging behind in technological development may streamline business processes and eliminate some inefficiencies in management or procurement, but merger is not equivalent to coming up to the modern of technological development. The evidence from the history of EDC shows that mergers and acquisitions work better if it starts with a company that has already been optimized and follows modern standards of efficiency.

The history of TGT points to the importance of specialization, as in the case of EDC. It is difficult to build a modern industry from scratch and enter a market that had already been divided by internationally recognized service providers. This barrier becomes less demanding in the case of a specific market niche where a start-up may have a definitive competitive (or comparative) advantage. The need of a specific competitive advantage leads to further specialization within a market niche where capitalizing on a unique technological advantage opens access to foreign markets and, therefore, an extended client base.

The filter of competitive market, once passed, guarantees that the company's core competence can become a basis for an expansion (if the company chooses the expansion strategy). Energy performance contracting in Russia is another example of a competitive market where companies constantly have to go through reality check.

The history of energy performance contracts in Russia is rather short. The regulatory environment for these services was established with the adoption of the Federal Law No. 261 dated November 23, 2009, "On Energy Saving and Development of Energy Efficiency and Amendments to the Russian

Federation Federal Laws" that triggered the development of performance contracts. This process was rather slow and came over several barriers including monopolization (Russian monopolies are often reluctant to deal with independent contractors especially small newly founded technological service providers) and devaluation of the Russian currency (the cost of new imported equipment, such as high-power energy efficient fluid pumps, has risen). The economic decline that the Russian Federation has been going through since 2014 should have made investments in energy efficiency more attractive, but EPC services themselves require investments, and more importantly companies may go bankrupt and default on their EPC during an economic decline. Turbulent economic conditions may cause unpredictable changes in tariffs and regulations that also contribute to the EPC risks (Garbuzova-Schlifter and Madlener, 2015).

Yet, there is a significant energy efficiency growth potential that is explained by the low base effect: researchers and government agencies confirm that Russian industry, commercial, and real estate sectors have considerable potential for energy efficiency enhancements (IFC, the World Bank, 2014; Zhang, 2011). It is equally often found that much of this potential has not been realized so far (Ministry of Energy of the Russian Federation, 2013). Many Russian energy systems, especially built during the Soviet era, were not meant to be energy efficient. The Soviet command economy provided no incentives for state R&D institutes to adhere to energy efficiency to the extent that some large-scale hydroelectric power plants may have been built without a proper engineering economic feasibility study (Kirillin, 1990).

Not all engineering or economic inefficiencies of Soviet industrial or energy systems can be eliminated within the frameworks of energy performance contracts that are supposed to make incremental modifications of an energy system, nor radical change of the system. For example, the economic feasibility study of district heating systems included the concept of a district heating radius* as a spatial extension at which the centralization of district heating still demonstrates a positive economy of scale when compared with distributed heat sources.

Despite such problems, EPC business slowly develops in Russia. The reason for that lies in its nature: EPC projects are analogous to financial arbitrage at the fundamental level in the sense that such projects capitalize on inefficiency. Then, the EPC market in Russia is open. Any organization specializing in repair, civil engineering, maintenance, or similar fields can become an EPC market agent. And there are many organizations whose machinery, technological processes, and buildings still have room for energy efficiency improvements. Thus, there is a potentially large demand. The entry barrier for EPC contractors is low. The average scale of a typical EPC project is rather small, such as installing LED lighting instead of old filament lamps. This combination can potentially make this market very competitive in the

---

* These can be described as a geographic measure of monopoly extension.

future, but at present the EPC market in Russia is still underdeveloped and competition is limited. The participation of foreign companies in the EPC market (so important in oilfield services, as shown in the following text) is moderate. There are Russian branches of international service companies, such as EDF Fenice, but their presence does not create intense competition. As a result, energy performance contracting is slowly taking off in Russia.*

3.5 Conclusion

It was shown that mergers and acquisitions work differently depending on the maturity and efficiency of companies. A merger or acquisition can potentially provide significant benefits for a cost-efficient business, but it may also result in a significant decrease of efficiency if the basic level was suboptimal.

In the Russian oilfield service sector and petroleum exploration, the situation is far from serene, but companies demonstrate contrasting trends. Those having chosen to specialize within the sphere of their competitive advantage naturally become innovative. Companies or organizations that have been merged instead of optimization get stuck with accumulated inefficiency. In some cases, uniting patchy assets in a larger company could potentially lead to an advantageous position (at least in the local market), but it does not spur innovation. More often the resulting corporations try to create an entry barrier for other companies or call for state support (quotas, tax cuts, etc.).

The reality check of competitive advantage implies that the company faces competition. The TGT case shows that companies surviving in the competition tend to prioritize innovation. This testifies to the pressing need to restructure the Russian oil and gas industry. Ideally, the reform should bring more competition into the industry and incentivize companies to develop new efficient technologies. This restructuring should ideally have been conducted during a more favorable period when soaring petroleum prices could attract investors. That opportunity was lost. It is also obvious that various types of subsidies to the industry do not work toward this cause because without restructuring they compensate for management inefficiency of Russian energy companies.

Basic energy research in the interest of the entire sector is necessary and some of it has been planned by several Russian Technology Platforms (Proskuryakova et al., 2016), but as in the case with subsidies the effect may be moderate. Restructuring is a painful process, especially when it is conducted

---

* Official estimates and forecasts are not always accurate: in 2011 the official forecasts of the EPC market were RUB 500 bln (Voskresensky, 2011), but the actual volume in 2015 was around 1% of this amount (RBC, 2015).

in harsh market conditions that do not seem to change soon, but it appears more and more necessary for the Russian energy sector.

Acknowledgment

This chapter was prepared within the framework of the Basic Research Program at the National Research University Higher School of Economics (HSE) and supported within the framework of a subsidy by the Russian Academic Excellence Project "5-100."

References

Agénor P.-R., Canuto O. (2015) Middle-income growth traps. *Research in Economics* 69(4), 641–660.

Akman U., Okay E., Okay N. (2013) Current snapshot of the Turkish ESCO market. *Energy Policy* 60, 106–115.

Azomahou T.T., Boucekkine R., Nguyen-Van P. (2008) Promoting clean technologies: The energy market structure crucially matters. UNU-MERIT Working Papers 2008/15, Maastricht, The Netherlands.

Banerjee D.K. (2012) *Oil Sands, Heavy Oil, and Bitumen: From Recovery to Refinery.* PennWell, Tulsa, OH.

Beicip Franlab (2016) Basin modelling and petroleum system analysis, available at: http://www.beicip.com/basin-modeling-and-petroleum-system-analysis (last accessed June 6, 2016).

Bertoldi P., Rezessy S., Vine E. (2006) Energy service companies in European countries: Current status and a strategy to foster their development. *Energy Policy* 34, 1818–1832.

Borge-Diez D., Colmenar-Santos A., Pérez-Molina C., López-Rey Á. (2015) Geothermal source heat pumps under energy services companies finance scheme to increase energy efficiency and production in stockbreeding facilities. *Energy* 88, 821–836.

Bustos F., Lazo C., Contreras J., Fuentes A. (2016) Analysis of a solar and aerothermal plant combined with a conventional system in an ESCO model in Chile. *Renewable and Sustainable Energy Reviews* 60, 1156–1167.

Dai X., Cheng L. (2016) Market distortions and aggregate productivity: Evidence from Chinese energy enterprises. *Energy Policy* 95, 304–313.

Dangelico R.M., Pontrandolfo P. (2015) Being "green and competitive": The impact of environmental actions and collaborations on firm performance. *Bus Strategy and the Environment* 24(6), 413–430.

Deloitte (2015) Russian oilfield service market: Current state and trends, available at: http://www2.deloitte.com/ru/en/pages/energy-and-resources/articles/oil-service-market-in-Russia-2015.html (last accessed June 6, 2016).

Deloitte (2016) Russian Oil & Gas Outlook Survey, 10th anniversary edition, Moscow, Russia, http://www2.deloitte.com/ru/en/pages/energy-and-resources/articles/russian-oil-gas-outlook-survey.html (last accessed May 6, 2016).

Eurasia Drilling Company Ltd. (2014) EDC land drilling fleet, available at: http://www.eurasiadrilling.com/operations/rig-fleet/edc-land-drilling-fleet/ (last accessed June 7, 2016).

Eurasia Drilling Company Ltd. (2015a) Completion of Merger and Notice of DeListing, available at: http://www.eurasiadrilling.com/media-centre/news-releases/2015/11/17/completion-of-merger-and-notice-of-delisting/ (last accessed June 6, 2016).

Eurasia Drilling Company Ltd. (2015b) Interim results, September 2015, http://www.eurasiadrilling.com/media/105486/EDC%20Presentation%202015%20Interim%20Results%20(September).pdf (last accessed January 5, 2017).

Fais B., Sabio N., Strachan N. (2016) The critical role of the industrial sector in reaching long-term emission reduction, energy efficiency and renewable targets. *Applied Energy* 162, 699–712.

Garbuzova-Schlifter M., Madlener R. (2015) Risk analysis of energy performance contracting projects in Russia. FCN Working Paper No. 10/2014. Institute for Future Energy Consumer Needs and Behavior, Aachen University, Aachen, Germany.

Gokhberg L. et al. (Eds.) (2016) *Indicators of Innovation in the Russian Federation 2016: Data Book*. National Research University Higher School of Economics, Moscow, Russia.

Hannon M., Foxon T., Gale W. (2013) The co-evolutionary relationship between Energy Service Companies and the UK energy system: Implications for a low-carbon transition. *Energy Policy* 61, 1031–1045.

Hannon M.J., Bolton R. (2015) UK Local Authority engagement with the Energy Service Company (ESCo) model: Key characteristics, benefits, limitations and considerations. *Energy Policy* 78, 198–212.

Hantschel T., Kauerauf A.I. (2009) *Fundamentals of Basin and Petroleum Systems Modeling*. Springer, Berlin, Germany.

Hirst E., Brown M.A. (1990) Closing the efficiency gap: Barriers to the efficient use of energy. *Resources, Conservation and Recycling* 3, 267–281.

IFC, the World Bank (2014) Energy efficiency in Russia: Untapped reserves, available at: http://www.cenef.ru/file/FINAL_EE_report_rus.pdf (last accessed May 25, 2016).

Ismail-Zadeh A., Korotkii A., Tsepelev I. (2016) *Data-Driven Numerical Modelling in Geodynamics: Methods and Applications*. Springer, Berlin, Germany.

Jaffe A., Stavins R. (1994) The energy-efficiency gap: What does it mean? *Energy Policy* 22, 804–810.

Kibsgaard P. 2015 *Scotia Howard Weil Energy Conference*, New Orleans, LA. Schlumberger, March 23, 2015, available at: http://www.slb.com/news/presentations/2015/~/media/Files/news/presentations/2015/Kibsgaard_Scotia_Weil_03232015.ashx (last accessed June 7, 2016).

Kindström D., Ottosson M. (2016) Local and regional energy companies offering energy services: Key activities and implications for the business model. *Applied Energy* 171, 491–500.

Kirillin V.A. (1990) *Key Problems (in Questions and Answers). The Tribune for an Academician*. Zhanie, Moscow, Russia (in Russian).

Kovalev A., Proskuryakova L. (2014) Innovation in Russian district heating: Opportunities, barriers, mechanisms. *Foresight and STI Governance* 8(3), 42–57.

Li J., Yu L. (2016) Double externalities, market structure and performance: An empirical study of Chinese unrenewable resource industries. *Journal of Cleaner Production* 126, 299–307.

Lombardi M., Pazienza P., Rana R. (2016) The EU environmental-energy policy for urban areas: The Covenant of Mayors, the ELENA program and the role of ESCos. *Energy Policy* 93, 33–40.

Lutzenhiser L., Biggart N.W., Kunkle R., Beamish T.D., Burr T. (2001) Market structure and energy efficiency: The case of new commercial buildings, available at: http://www.uc-ciee.org/downloads/market_struc.pdf (last accessed June 6, 2016).

Mahapatra K., Gustavsson L., Haavik T., Aabrekk S., Svendsen S. et al. (2013) Business models for full service energy renovation of single family houses in Nordic countries. *Applied Energy* 112, 1558–1565.

Malysheva S. (2015) Regional modelling of basins of various geodynamic types in connection with their forecasted oil and gas bearing capacity. Candidate thesis dissertation. Gazpromneft Science and Technology Center, St. Petersburg, Russia, available at: http://www.vnigni.ru/diss_s/malysheva/disser.pdf (last accessed June 6, 2016).

Ministry of Energy of the Russian Federation (2013) Energy efficiency policy in Russia, available at: http://www.unescap.org/sites/default/files/D_Russia_NadezhdinEEPresentation_1.pdf (last accessed February 25, 2016).

Oklahoma Corporation Commission (2014) Available at: http://imaging.occeweb.com/OG/Well%20Records/00000005/OCC_OG_0G5BFM7_1L8M2C4.pdf (last accessed May 30, 2016).

Pacheco M.A.C., Vellasco M.M.B.R. (Eds.) (2009) *Intelligent Systems in Oil Field Development under Uncertainty*. Springer, Berlin, Germany.

Painuly J.P., Park H., Lee M.K., Noh J. (2003) Promoting energy efficiency financing and ESCos in developing countries: Mechanisms and barriers. *Journal of Cleaner Production* 11, 659–665.

Pätäri S., Sinkkonen K. (2014) Energy service companies and energy performance contracting: Is there a need to renew the business model? Insights from a Delphi study. *Journal of Cleaner Production* 66, 264–271.

Proskuryakova L., Filippov S. (2015) Energy technology Foresight 2030 in Russia: An outlook for safer and more efficient energy future. *Energy Procedia* 75, 2798–2806.

Proskuryakova L., Meissner D., Rudnik P. (2016) The use of technology platforms as a policy tool to address research challenges and technology transfer. *The Journal of Technology Transfer* 41(5), 1–22.

RBC (2015) Russian energy service market 2015. RBC Research (in Russian), Moscow, available at: http://marketing.rbc.ru/research/562949998570476.shtml (last accessed June 6, 2016).

Reis J.C. (1996) *Environmental Control in Petroleum Engineering*. Gulf Publishing, Houston, TX.

Rosgeologiya (2016) Rosgeologiya will access the prospects of oil and gas bearing capacity of Udmurtiya region, available at: http://www.rosgeo.com/ru/content/rosgeologiya-ocenit-perspektivy-neftegazonosnosti-udmurtskoy-respubliki (last accessed June 6, 2016).

Rosneft (2015) Sakhalin-1 sets another extended reach drilling record, available at. https://www.rosneft.com/press/today/item/175093/ (last access: 05.01.2017).

Schlumberger (2016) PetroMod petroleum systems modeling software, available at: https://www.software.slb.com/products/petromod (last accessed June 6, 2016).

Soderholm P., Warell L. (2011) Market opening and third party access in district heating networks. *Energy Policy* 39, 742–752.

Suhonen N., Okkonen L. (2013) The Energy Services Company (ESCo) as business model for heat entrepreneurship—A case study of North Karelia, Finland. *Energy Policy* 61, 783–787.

Tretiyakov (2014) Oil companies should drill more in order to sustain the volumes of oil extraction. *Vedomosti* 3702, 23.10.2014 (in Russian), available at: https://www.vedomosti.ru/business/articles/2014/10/23/neftyanikam-nado-bolshe-burit (last accessed June 7, 2016).

U.S. National Science and Technology Council (2011) Materials Genome Initiative for Global Competitiveness, available at: https://www.whitehouse.gov/sites/default/files/microsites/ostp/materials_genome_initiative-final.pdf (last accessed June 6, 2016).

Vine E. (2005) An international survey of the energy service company ESCO industry. *Energy Policy* 33, 691–704.

Voskresensky S. (2011) XVIII Krasnoyarsk Economic Forum, February 17–19, 2011.

Wangen M. (2010) *Physical Principles of Sedimentary Basin Analysis*. Cambridge University Press, Cambridge.

Worrell E., Laitner J., Ruth M., Finman H. (2003) Productivity benefits of industrial energy efficiency measures. *Energy* 28, 1081–1098.

Zhang Y.-J. (2011) Interpreting the dynamic nexus between energy consumption and economic growth: Empirical evidence from Russia. *Energy Policy* 39(5), 2265–2272.

Zirax Nefteservice (2016) Corporate website. http://www.zirax-nefteservice.ru/ (last accessed May 28, 2016).

4

Competitiveness of Distributed Generation of Heat, Power, and Cooling: System Design and Policy Overview

Konstantinos C. Kavvadias

CONTENTS

4.1 Introduction

4.1.1 Concept and Definition

Distributed generation (DG) of energy is not a new concept, but it has been showing an ever-increasing promise as a cost-effective and energy-efficient method of energy supply. While there are some attempts for a concrete definition of DG, there is no universally accepted one. Many definitions focus on distinguishing them from large centralized generation units. The key concept lies in its local nature: small-scale energy generation units located at or close to the place where the energy products are consumed, thus bypassing the transfer and distribution network. It is part of the *distributed energy resources* paradigm, which also contains demand response and distributed storage.

Many definitions that identify distributed energy generation technologies are based on size (Ackermann et al. 2001). According to these, there is a specific threshold, usually from 0.5 to 2 MW, under which a technology can be characterized as DG. Others assume that DG units are placed at or near customer sites to meet specific customer needs or to support economic operation of the grid, or both (Pepermans et al. 2005). However, the size is not the only criterion that defines the concept. DG also has the following characteristics:

- *Technology*: It can be implemented with different technologies that are efficient in small scale, easy to deploy usually as integrated energy solutions, having a reasonable footprint. It can also be either renewable or nonrenewable.
- *Ownership*: Such installations can be owned either by the end user, a utility, or an energy service company.
- *Environment*: The increased efficiency and advanced technology usually result in lower emissions. There are some concerns that decentralizing the emissions will not always have a positive effect as it can disproportionally burden the atmospheric conditions of urban areas.
- *Purpose*: The consumer can either exclusively use the energy produced or sell it partially or wholly to the grid/heat network according to the regulatory framework. The equipment can be completely stand-alone or connected to a grid. In any case, DG affects grid operation either by lowering its demand (since the energy is produced on-site) or by infusing new energy.
- *Energy products*: Lately, DG has been viewed from a more systemic perspective, as it deals with multiple energy vectors (distributed multigeneration), referring not only to multiple energy carriers (e.g., electricity, hydrogen) but also to multiple energy uses like heating and cooling (Lund et al. 2012; Mancarella 2014).

- *Dispatch strategy*: The dispatch strategy depends on the flexibility of the technology. In that respect, there are three types of power production: intermittent, constant, and flexible power generators. The first one depends on the availability of the resource (e.g., wind, sun) and can be dispatched only when it is available. Fossil fuel–based technologies can either operate continuously or respond to price evolution, thus serving as a hedge to abrupt price fluctuations (e.g., during peak time).

From these characteristics, it is clear that DG refers to different aspects and technologies, which will be covered in the following sections.

4.1.2 Centralized versus Distributed Generation of Heat, Power, and Cooling

Traditionally, electricity was produced centrally in big plants and heat was produced at the place of the demand. Technical reasons were driving this pattern: electricity generation was mainly based on thermodynamic cycles (e.g., Rankine, Brayton), which performed better on large scales; scaling down these plants is a technological challenge, which costs a lot. Electricity is a convenient energy carrier, which can be transferred and distributed easily, so there were large investments in transmission and distribution networks.

Technology improvements have allowed power generation systems to be built on-site where the energy is going to be used without sacrificing efficiency, cost, or environmental impact. In order to maximize the overall energy efficiency, the generation units were modified so that the waste heat could be captured, covering more end uses such as heating and cooling. Cogeneration of heat and power was deployed at a fast pace; later, trigeneration (Kavvadias et al. 2010)—which is another way to describe the combined generation of heat, power, and cooling—and currently multigeneration of different energy carriers by different primary fuels have been rising in popularity among researchers (Chicco and Mancarella 2009; Mancarella 2014). These generation units, coupled with smart grids, which allow connection of active consumers and prosumers and also monitor and repair themselves, can shape the future of electricity generation and transmission network.

On the other hand, heat (and cooling) has been traditionally produced on-site and as a result its production was already decentralized. Heat as a commodity is more difficult to handle than electricity because it has to be transferred physically with a medium (usually water) compared to electricity, which only has electron movement. The concept of harnessing waste heat dissipated by the thermodynamic cycle had made its appearance in the centralized plants, but in this case, pipe networks had to be built to transfer the heat to nearby consumers. The original networks were steam based but nowadays they are water based, with a trend to lower their operating temperature, which allows a more flexible operation, in the same way as smart grids are revolutionizing electricity networks (Lund et al. 2014).

Thus, there was an expansion in the scope of both centralized and decentralized generation of energy to multigeneration:

1. Centralized power plants extended their primary activity of electricity production to heat production and distributed it via other network to the end consumers (main producers).
2. DG plants were built by various consumers generating electricity and heat wholly or partly for their own needs, an activity that supported their primary activity (autoproducers).

Both concepts emerged from the need for increased energy efficiency. At first glance, the fact that electricity production is decentralized and heat production is getting centralized may seem like a paradox. However, both events are driven by the need for increased overall energy efficiency. When properly planned, both can lead to an overall energy efficiency increase of the energy systems and to a more flexible and diversified energy system with an increased energy supply.

4.1.3 Benefits of Distributed Generation

Such installations bring significant benefits to consumers and to the electricity grid. The main categories are summarized here (Colmenar-Santos et al. 2016).

4.1.3.1 Technical

- *Power loss reduction*: Generation closer to demand, which prevents loss of energy in transmission networks.
- *Improvement of energy efficiency*: The use of combined heating and power (CHP) units allows the simultaneous generation of heat and electricity and hence improvement in energy efficiency average of the system.
- *Improvement of security supply*: The increasing penetration of DG and future intelligent networks (smartgrid) can contribute to increasing security supply, as it will diversify the primary energy supplies and potentially reduce the dependency for foreign sources.
- *Improvement of voltage profiles*: Connection of DG to a network enables normal raise in voltage, which can contribute to an improved voltage profile, especially in radial distribution networks in medium and low voltage.
- *Increase of quality power*: In areas where voltage levels are low, installation of DG can improve the quality of supply.

4.1.3.2 Economic

- *Reduction of operative cost*: Cost reduction in transmission and distribution of energy, hence reduction in losses; and reduction in maintenance costs (failures and lines congestion).
- *Reduction of capital costs*: DG can delay the need for investments in new transmission and distribution infrastructures and reduce depreciation costs of fixed assets in networks.
- *Reduction of environmental costs*: Reduction of emissions into the atmosphere helps to reduce associated costs with environmental penalties.
- *Electricity tariff reduction*: The increased penetration of DG can open energy markets to new agents and low prices.
- *Liberalization of energy markets*: By allowing market agents to install their own energy generation equipment they can respond to changing market conditions, increasing the flexibility of the system and promoting competitiveness, which can lower the overall prices.

4.1.3.3 Environmental

- *Reduction of fossil fuel consumption*: The use of distributed renewable sources and the increased efficiency of multigeneration plants reduce fossil fuel consumption in conventional power plants.
- *Reduction in greenhouse gas emission*: The reduction of fossil fuel consumption implies the reduction of SOx and NOx emissions into the atmosphere.

4.2 Distributed Generation Technologies

4.2.1 Distributed Combined Generation (Autoproducers)

For the reasons described earlier, one of the most popular DG technologies is CHP. It is recognized by the European Union (EU) as one of the most efficient ways to produce end-use energy from fossil fuels (EU 2004, 2012). CHP systems cover all five dimensions (energy efficiency, secure supplies, energy market, emission reduction, research, and innovation) of the EU's Energy Union and its heating and cooling strategy. The following sections describe the design consideration and competitiveness of such systems.

4.2.1.1 System Design

System design aims at determining the sizing and operational variables involved by optimizing a suitable criterion. The proposed design solution must be subject to the restrictions fixed by the legislation, while it is very sensitive to the country's energy policy and to wider geopolitical facts (i.e., abrupt oil price change). Optimization based on economic criteria from the investor's point of view has been studied thoroughly for both cogeneration and trigeneration plants.

The main principle of a multigeneration plant is that it converts fuel energy directly to mechanical shaft power, which can drive a generator to produce electricity. Waste heat can be recovered to cover the thermal demand or cooling demand via an absorption chiller. A conceptual energy flow diagram of a trigeneration plant is presented in Figure 4.1.

Two prime mover types that are most popular for DG applications are described in the following subsections.

4.2.1.1.1 Internal Combustion Engine

An internal combustion engine (ICE) is the most mature prime mover technology used in distributed co/trigeneration systems and is mainly driven by natural gas in spark-ignition engines. Compression-ignition engines can also run on diesel fuel or heavy oil. Reciprocating engines are a proven technology with a range of sizes and the lowest capital costs of all combined cooling, heating, and power (CCHP) systems. In addition to fast start-up capability and good operating reliability, high efficiency at partial load operation give users a flexible power source, allowing for a range of different energy applications—especially emergency or standby power supplies. Moreover, they have relatively high electrical efficiency (35%). Reciprocating engines are by far the

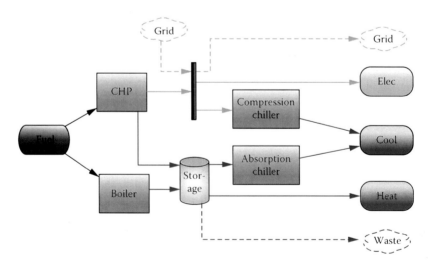

FIGURE 4.1
Conceptual energy flow diagram of a trigeneration plant.

most commonly used power generation equipment between 100 and 5000 kW, because they have an almost flat efficiency curve above 30% of the nominal electrical power (Badami et al. 2008). This implies that they can work success-fully on part loads and several operation strategies can be applied successfully.

4.2.1.1.2 Micro-Turbines

Micro-turbines (µT) are actually an extension of turbine technology on a smaller scale. They are primarily fueled with natural gas, but they can also operate with diesel, gasoline, or other similar high-energy fuels. Research on biogas is ongoing. µTs have only one moving part; they use air bearings and they do not need lubricating oil, although they have extremely high rotational speed, up to 120,000 rpm. Small-scale individual units offer great flexibil-ity and can be easily combined into large systems of multiple units. During their operation they have low noise and relatively low NOx emissions. On the other hand, they usually have low electricity efficiency and high cost.

Other prime movers are also used in CHP such as fuel cells or Stirling engines, but they are still under development and their economics are not very favorable for commercial applications.

Following is the description of an ICE-driven trigeneration plant: An inter-nal combustion engine is fed by air and natural gas as fuel. Through the com-bustion process, the chemical energy of the fuel is converted into mechanical shaft power, which drives a generator to produce electricity. ICEs operate either according to the Otto or diesel cycle and waste heat is generated at two temperature levels: by a low-temperature flow of coolant (90°C–125°C) and by a medium-temperature flow of exhaust gas (200°C–400°C). The ICE exhaust gases can be used either directly in thermal processes or indirectly through a heat recovery steam generator, which produces superheated steam. Most auto-producers do not need very high-grade heat, so it is assumed that there is no need to use the exhaust gases directly. When the thermal output of the prime mover is not sufficient to cover the demand, a boiler is required to operate.

In a trigeneration plant, cooling energy can be generated in two ways: either by utilizing waste heat via an absorption chiller or by utilizing electric-ity via an electric heat pump. Electric chillers use a mechanical compressor in order to take the refrigerant vapor from the lower evaporation pressure to the higher condensation pressure. In absorption chillers, this process is real-ized by means of a solution circuit, which serves as a thermal compressor. Absorption chiller cycles are based on certain thermodynamic properties of two fluids: one is the refrigerant and the other is the absorbent. The most common pairs found in the literature are as follows:

- Ammonia as the refrigerant and water as the absorbent. Such a com-bination is chosen when low evaporation temperatures are needed (below 0°C).
- Water as the refrigerant and a solution of lithium bromide as the absorbent.

Vapor generated in the evaporator is absorbed into the liquid absorbent in the absorber. The absorber that has taken up the refrigerant is pumped into the generator. The refrigerant is released again as vapor by waste heat from steam (or hot water) is to be condensed in the condenser. The regenerated absorbent is led back to the absorber to pick up the refrigerant vapor (Wu and Wang 2006).

In general, absorption chillers are fueled by the exhaust thermal energy from the prime mover. This reduces peak electric demand and electricity charges by reducing the operating time of electric chillers and increasing the electric to thermal load coincidence in the summer months. It must be mentioned, though, that when waste heat is not available it is not always economically viable to generate heat by burning fuel due to the small coefficient of performance (COP) of the absorption chiller (0.7–1.2) compared to that of the electrical one (2.5–5). Hence, the absorption chiller should be preferred over the electrical chiller only when waste heat is available, or when cooling demand is significantly bigger than heating demand. The main differences between these two technologies are summarized in Table 4.1.

Finally, a buffer vessel is utilized in order to balance the hourly fluctuations of thermal demand, which will accumulate the heat produced that is not needed at a specific moment, consequently "smoothing" the peaks. Energy is stored at times when the available means of generation exceed demand and is returned when demand exceeds supply. Thermal energy is, in practice, the only form of energy that can be stored by consumers. Electricity can be stored locally, but at a much higher cost than storage at the supply side, and is in most cases not efficient. For this reason, a storage system for electricity (i.e., battery) is not examined.

TABLE 4.1

Comparison of Absorption Chillers with Conventional Vapor Compression Chillers

	Vapor Compression	Absorption Chiller
Energy source	Electricity	Heat
Part load behavior	Medium	Very good
Mechanical moving parts	Many	Few
Maintenance costs	High	Low
Investment costs	Low	Low
Coefficient of performance (COP)	High	Low
Water consumption in cooling tower	Medium	High
Unit weight	Medium	Big
Noise vibration	Medium	Low
Greenhouse gases in coolant liquid	Yes	No

4.2.1.2 Selection and Design Considerations of a
Distributed Cogeneration Plant

Selecting and sizing a CHP plant depends on the heat and electricity energy demand and on the coincidence of these loads. Heat-to-power ratio is one of the most important technical parameters influencing the selection of the type of cogeneration system. The heat-to-power ratio of a facility has to match with the characteristics of the cogeneration system to be installed.

The design and sizing of a plant based on the load duration curve of the heat demand is a commonly used rule of thumb subject to several limitations such as lack of load coincidence information, assumption of ideal operation, etc. (Piacentino and Cardona 2008). Some sector-specific considerations for the design of CHP are presented here.

Industry: Industry loads are usually easy to predict and simulate because they depend heavily on the production needs and planning. Energy is used mainly in production processes and less in heating, ventilating, and air-conditioning (HVAC) or lightning systems. For this reason, ambient temperature has a smaller effect. Industries can be divided into two big categories in regard to their type of operation: continuous or batch. Those that have continuous operation have more constant loads, and the fraction between the energy kinds does not fluctuate a lot during the day. The latter depend only on the kind of industry. It is clear that industries can be classified according to their energy priorities, for example, an aluminum industry is very power-intensive whereas an ice-cream industry needs large quantities of cooling energy. A very common problem that occurs in batch operation industries is the successive alternation of energy demand; for example, a heat-intensive process is needed immediately after a power-intensive process is completed, thus making these industries unsuitable for CHP systems.

Commercial buildings have demanding thermal and cooling loads due to HVAC systems. For this reason, CHP technology and application matching in the commercial sector is more difficult than industrial complexes as (1) it has more fuzzy profiles, (2) on average it operates fewer hours per year so the payback period of the investment rises, and (3) it is generally smaller than industrial sites, which means it is less efficient and has smaller economy scales. In contrast to industrial consumers, ambient temperature heavily affects commercial buildings. Of course, the most important factor is occupation and activity frequency. Both seasonal and daily variations of energy need to be considered for a more precise design result. Electricity is usually distributed to office applications and cooling devices. Thermal energy is used for space heating and other processes that need general heating, such as equipment sterilization, laundry, and kitchen.

Residential buildings have the most unpredictable loads since they are based on human acts and needs. The cyclic variance of energy demand due to the operating nature of residential equipment (fridge, boiler, etc.) is important to consider even on a half-hour basis. The most important factors that affect occupation pattern and thus the energy demand are (Yao and Steemers 2005) (1) the number of residents, (2) the time during which the first resident stands and goes to sleep, and (3) the time that a house is unoccupied during the day.

4.2.1.3 Operation Strategies

Operation strategies that are used in DG plants are part of the process control system, which is dependent on the following factors: demand for each kind of energy, prime mover nominal power, coefficients of performance, and conversion factors for all energy conversion devices involved. In the literature, the most common kinds of cogeneration systems are designed by either covering a constant part of energy or by following the evolution of the electrical (or heat) load.

The following operation strategies can be identified for multigeneration units (Kavvadias et al. 2010):

1. *Continuous operation*: The system operates on maximum power. This strategy can be used in order to cover the base load. An auxiliary boiler produces thermal energy when needed to cover the heat load. If a bigger prime mover is utilized, the excess electricity can be sold to the grid.

2. *Peak saving*: The system operates for a limited amount of time to cover a predefined part of the load during electricity peak conditions. As a result, the peak power bought from the grid is reduced or the utilization factor is improved resulting in cheaper marginal electricity prices.

3. *Electricity equivalent demand following*: The system operates in order to cover the electricity load and the electricity needed for the electric chiller minus the electricity that is conserved by the operation of the absorber in order to cover the cooling load. Thermal energy is produced (via an auxiliary boiler) or wasted in order to integrate the rest of the energy demand or offer, respectively.

4. *Heat equivalent demand following*: The system operates in order to cover the heat load and the heat needed for the absorption chiller to cover the cooling load. Electricity is bought from or sold to the grid in order to integrate the rest of the energy demand or offer, respectively.

4.2.1.4 Competitiveness of Distributed Generation

In order to better understand what affects the competitiveness, it is necessary to examine the investment initiative of autoproducers as defined in Section 4.1.2. The investment dilemma of autoproducers consists of the decision as to whether cogeneration is more economical than conventional outsourced separate generation production means. The driving force of CHP investments is the energy savings, and the profits related to these savings are linked mainly to the prices of the competing fuels, which are usually gas and electricity. It is evident that the more efficient the substituted equipment, the less attractive the energy efficiency investment is going to be. Another driver for DG is the displacement of high-priced grid power with lower-cost electricity generated on-site. Project economics are based on many project-specific factors: size of the system, total installed cost of the project, and local construction and labor rates. Production of energy is not the core business of the autoproducers, so a stable and risk-free environment is needed. In other words, these consumers (especially from the commercial and residential sectors) show preference to systems that are simpler and not as price-inelastic as cogeneration systems (Lončar et al. 2009). Hence, the competitiveness of such installations is dependent on the substitution of the current equipment, market conditions, and the stability that is provided by the regulatory framework.

Accordingly, it makes sense to study the theoretical relation between the viability of combined generation technologies and the market conditions and conventional equipment efficiency. "Spark spread," which refers to either the difference or the ratio of the competitive fuels, that is, natural gas and electricity, is the most common indicator. In an "energy market" context, it is usually the difference between electricity prices and gas prices multiplied by the heat rate, which reflects the gross operation margin of a power plant (Sezgen et al. 2007; Wright et al. 2014). Based on this difference, many financial products or options have been used to hedge (Palzer et al. 2013) and to estimate the value of such investments (Mancarella 2014; Sezgen et al. 2007).

Dispatch decisions between competing technologies (e.g., cogeneration vs. heat pumps) have also been based on this difference (Capuder and Mancarella 2014). For CHP to be profitable, the U.S. Department of Energy Midwest CHP Application Center (2007) proposes at least a difference of $0.04/kWh between natural gas and electricity. This rule of thumb refers only to CHP prime movers and does not consider the characteristics of substituted conventional equipment. However, other reports use the price ratio to identify the feasibility of CHP. A latest report on European cogeneration (Cogeneration Observatory and Dissemination Europe 2015) states that the ratio between electricity and fuel prices should be around 3 without any further justification and link to specific equipment. Cardona et al. (2006a,b) used this price ratio to develop an operation strategy which, on an hourly basis, can decide whether a CHP prime mover should operate or not. Graves et al. (2008) developed a method that correlates the prime mover efficiency, the heat recovery ratio, and the equipment cost as an indication of CHP viability. Smith et al. (2011a,b) developed a similar indicator that is based on the operational characteristics of CHP but did not generalize it for the case of cooling production.

The literature review does not conclude with a generic feasibility indicator that correlates the energy prices with specific cogeneration technologies independent of the energy loads. Such indicators are being used extensively, but as discussed in the previous paragraph, the choice of values is governed by empiricism having limited applicability. The development of a theoretical relation between energy prices and the characteristics of cogeneration and conventional generation equipment is made in the next subsection.

4.2.1.5 Theoretical Formulation of Competitiveness

In this subsection, there will be an attempt to map the operational viability of co- and trigeneration equipment and to give a clear view of the sensitivity of energy prices on energy efficiency investments.

Figure 4.2 shows the reference energy system that will be used for this study. An energy consumer demands three energy products (electricity, heating, and cooling) at any given time. These loads can be covered in the following ways: either via combined generation (left side of Figure 4.2) or via conventional generation (right side of Figure 4.2) technologies. The combined generation system consists of a prime mover (internal combustion engine, gas turbine, etc.) with a heat recovery system and a thermal-driven heat pump, such as an absorption chiller, which utilizes low-grade heat. The conventional generation part consists of grid electricity, a fossil-fueled boiler, and an electric-driven heat pump.

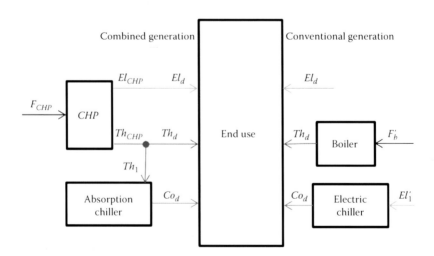

FIGURE 4.2
Reference energy system for the coverage of specific energy demand by cogeneration and conventional generation. (Adapted from Kavvadias, K.C., *Energy*, 115(3), 1632–1639, 2016.)

In the context of this comparison, the energy that is covered by other sources is ignored (e.g., grid, boiler, and electric chiller) and the energy that can be produced by a CHP system (with predefined technical characteristics) is compared for a given time frame.

This system can be mathematically formulated as follows: let El_d (kW), Th_d (kW), and Co_d (kW) be the energy demand for electricity, heating, and cooling of an individual consumer, respectively.

For the *CHP* part:

$$Th_{CHP} = Th_d + Th_1 \tag{4.1}$$

$$Co_d = COP_{ab} \cdot Th_1 \tag{4.2}$$

where COP_{ab} is the coefficient of performance of the absorption chiller.

From these equations and the definition of the overall *CHP* efficiency, the following is derived:

$$\eta_{CHP} = \frac{El_{CHP}}{F_{CHP}} + \frac{Th_{CHP}}{F_{CHP}} \tag{4.3}$$

Similarly for conventional generation:

$$Co_d = COP_{El} \cdot El_1' \tag{4.4}$$

$$Th_d = \eta_b \cdot F_b' \tag{4.5}$$

The hourly operational cost of trigeneration is defined by means of

$$C_{CHP} = C_f \cdot F_{CHP} \tag{4.6}$$

whereas for conventional (separate) generation it is calculated by means of

$$C_{SHP} = C_f \cdot F_b' + C_e \cdot \left(El_d + El_1' \right) \tag{4.7}$$

where
  C_f (EUR/kWh) is the fuel price
  C_e (EUR/kWh) is the electricity price

A necessary assumption to be made is that the electricity and fuel costs for both conventional and CHP generation of energy are the same. This may not be the case if special policies and subsidies are applied, but it is useful to compare the inherent advantages and the true competitiveness of the two technologies.

As mentioned earlier, the basic investment motivation can be summarized as follows: when heating and electricity can be locally produced at a smaller cost than the grid electricity and separate heat generation, then and only then a DG CHP investment can operate with a profit.

For an economically viable operation of a trigeneration installation, the operating cost of the CHP unit has to be less than or equal to the cost of the conventional generation part for given energy loads:

$$C_{SHP} - C_{CHP} \geq 0 \qquad (4.8)$$

where C (*EUR*) is the operating costs of combined generation (*CHP*) and conventional generation (*SHP*), respectively, as defined in Equations 4.6 and 4.7.

Using the expressions (4.6) and (4.7) and replacing F_{CHP} from (4.3), F_b' from (4.5), El_1' from (4.4), and Co_d from (4.2), Equation 4.8 becomes

$$C_f \cdot \frac{Th_{CHP} - Th_1}{\eta_b} + C_e \cdot \left(El_{CHP} + \frac{Th_1 \cdot COP_{ab}}{COP_{el}} \right) - C_f \cdot \frac{El_{CHP} + Th_{CHP}}{\eta_{CHP}} \geq 0 \qquad (4.9)$$

We define the ratio of electricity to natural gas price as $PriceRatio = (C_e/C_f)$, the heat-to-power ratio of the prime mover as $HPR = (Th_{CHP}/El_{CHP})$, and the fraction of recovered heat that is used for cooling as $a = (Th_1/Th_{CHP})$. Replacing the variables in Equation 4.9 and dividing by Th_{CHP}, thus simplifying and solving the *PriceRatio*, the following equation is derived:

$$PriceRatio \geq \frac{COP_{el} \left[\eta_b + HPR \left(\eta_b - \eta_{CHP} + a \, \eta_{CHP} \right) \right]}{\eta_b \, \eta_{CHP} \left(COP_{el} + COP_{ab} \, HPR a \right)} \qquad (4.10)$$

For $\alpha = 1$, that is, when all heat is used for the production of cooling in the absorption chiller, the equation is simplified as follows:

$$PriceRatio \geq \frac{COP_{el} \left(1 + HPR \right)}{\eta_{CHP} \left(COP_{el} + COP_{ab} \, HPR \right)} \qquad (4.11)$$

whereas for $\alpha = 0$, that is, for simple cogeneration mode without an absorption chiller, the equation is simplified as follows:

$$PriceRatio \geq \frac{HPR + 1}{\eta_{CHP}} - \frac{HPR}{\eta_b} \qquad (4.12)$$

This relation covers only the operation feasibility ignoring the investment costs. Equation 4.8 can be modified so that it calculates operational costs on an annual basis including an annualized capital costs term:

$$\left(C_{SHP} - C_{CHP} \right) \cdot CapF \cdot 8760 - crf \left(l, i \right) \cdot \left(C_{eq\,CHP} \cdot El_d + C_{eq\,ab} \cdot Co_d \right) \geq 0 \qquad (4.13)$$

where

$C_{eq\,CHP}$ is the capital costs of a CHP unit (EUR/kW$_e$)

$C_{eq\,ab}$ (EUR/kW$_c$) is the capital costs of an absorption chiller

crf (—) is the capital recovery factor used to convert a present value into a stream of equal annual payments over a specified time (l), at a specified discount rate (i) by means of $crf = (i(1+i)^n/(1+i)^n - 1)$

$CapF$ (%) is the annual capacity factor of the cogeneration unit which is multiplied by 8760 (hours/year) to express the annual operating hours of the combined generation installation

This conversion is necessary for the dimensional consistency of the formula in order to express and compare all costs on an annual basis.

Equation 4.13 is solved in a similar way, but it cannot be simplified to a single price ratio due to an intercept term derived by the capital cost term. It will be shown in the results of the next section that the minimum gas price for a viable combined generation investment varies linearly as a function of electricity prices ($C_f < a \cdot C_e + b$).

4.2.1.6 Analysis of Spark Spread Sensitivity on Various Characteristics

Through the developed indicator the viability of different cogeneration technologies and configurations can be explored. The inequality (4.10) is not a function of the energy loads, but only a function of the technical specifications of the combined generation and conventional equipment. As it was mathematically proven, the operational viability is a function of the ratio and not the difference of the prices, as it is mentioned sometimes in the literature. Converting this inequality expression to equality, the operational breakeven point is estimated, that is, the price ratio for which CHP has the same operating costs as conventional generation. The most important innovation of the described generalized formulation is that the minimum price spread can now be mathematically justified based on given technical specifications and not on empiricism. The description of the inherent relationship of combined generation viability allows the system operators to regulate their CHP system and the decision-makers to quantify a minimum fuel subsidy in order to annihilate the operating risk of cogeneration units.

4.2.1.6.1 Operational Viability

In the following paragraphs, the effect of the equipment's technical specifications on the minimum *PriceRatio*, for which a combined generation system can operate profitably, is shown. Table 4.2 presents typical parameters of an internal combustion engine–based cogeneration unit. This type of unit is usually the ideal technology for middle-scale cogeneration systems used in buildings of the tertiary sector. Typical values of the conventional heating and cooling generation systems are also considered.

TABLE 4.2

Typical Values for Parameters of Equation 4.10

Parameters of Equation	Variable	Central Value
Coefficient of performance of electric chiller	COP_{el}	3.5
Coefficient of performance of absorption chiller	COP_{ab}	0.8
Boiler efficiency	η_b	85%
CHP overall efficiency	η_{CHP}	90%
Heat-to-power ratio of prime mover	HPR	1.2

Source: Adapted from Kavvadias, K.C., *Energy*, 2016.

The variable α can be used to simulate the seasonality effect of a combined generation device. During the summer months when a big percentage of heat is going to the absorption chiller, *a* tends toward 1. On the other hand, during the winter, α is usually 0 as all the recovered heat is directed for other end uses (space heating, hot water, etc.).

Figure 4.3 shows that the bigger the amount of heat that is used for cooling, the larger the *PriceRatio* has to be, that is, the natural gas price has to be much smaller than the electricity price. This correlation is explained due to the nonefficient conversion of heat in the one-stage absorption chiller ($COP < 1$). This means that during summer months when the needs for cooling are bigger, the need for cheaper natural gas is bigger. If this is not the case, then α has to be reduced by covering the cooling demand via other production means. This observation comes in line with what is applied in practice; the operation and installation of an absorption chiller are not viable beyond a specific natural gas price threshold.

The technical characteristics of the combined generation equipment positively affect the minimum price ratio, whereas the characteristics of the substituted equipment affect it negatively. The more efficient the new equipment

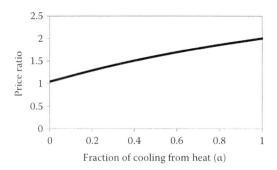

FIGURE 4.3

Effect of cooling fraction from recovered heat on the minimum PriceRatio. (Adapted from Kavvadias, K.C., *Energy*, 115(3), 1632, 2016.)

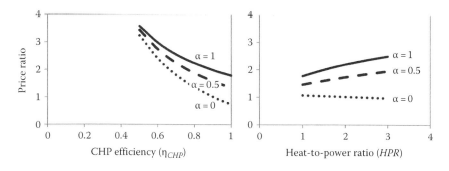

FIGURE 4.4
Effect of CHP prime mover characteristics on the minimum PriceRatio. (Adapted from Kavvadias, K.C., *Energy*, 115(3), 1632, 2016.)

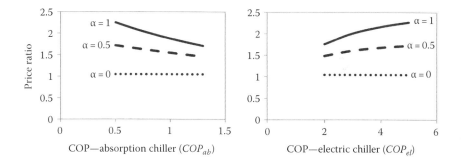

FIGURE 4.5
Effect of COP on the minimum PriceRatio. (Adapted from Kavvadias, K.C., *Energy*, 115(3), 1632, 2016.)

and the less efficient the substituted equipment, the smaller is the requirement for a high electricity to gas price ratio (Figure 4.4). Prime movers that produce more heat than electricity (for a given overall efficiency) are more sensitive to the variations of energy prices. For heating and electricity generation mode (no cooling), the heat-to-power ratio has a very small effect. Regarding cooling equipment, as expected, Figure 4.5 illustrates that the conventional and cogeneration equipment have an inverse relationship; the bigger the COP of the electric chiller and the lower the COP of the absorption chiller, the higher the minimum PriceRatio has to be.

The efficiency of the conventional boiler is apparently the most important variable (Figure 4.6), especially for operating conditions with small α (no cooling). If the equipment substitutes old nonefficient equipment, then the profit margin is very large. *PriceRatio* can even fall below 1, that is, CHP will be viable even if electricity prices are smaller than natural gas prices. In old and inefficient boilers, the CHP unit will be able to operate at any gas price, depending on the cooling fraction from heat as defined by α.

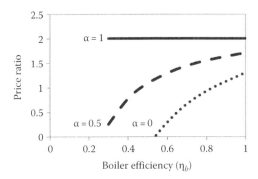

FIGURE 4.6
Effect of substituted boiler efficiency on the minimum PriceRatio. (Adapted from Kavvadias, K.C., *Energy*, 115(3), 1632, 2016.)

4.2.1.6.2 Investment Viability

The previous analysis was done for existing cogeneration devices. For new investments, the capital cost and the operating time of the equipment have to be calculated by means of Equation 4.13. In order to clarify the interactions between the critical variables and economic feasibility of new investments described in the previous sections, a simple sensitivity analysis is conducted. The plotted line in Figure 4.7 corresponds to the locus of the points where total annual costs (including depreciation of investment) of separate production are the same as in the cogeneration case (investment break-even line). In order to better illustrate the difference between the heating ($a = 0$) and cooling ($a = 1$) mode, two break-even charts were plotted. For the combination of prices that fall within the area above each line, separate production is more economical. In the area below the line, cogeneration is more economical. Three different prime movers are compared: internal combustion engine (ICE), gas turbine (GT), and micro-turbine (μT). The technology parameters assumed are presented in Table 4.3. Parameters from conventional equipment are adopted from Table 4.2. It has to be noted that a higher capacity factor applies to the heating and cooling modes (trigeneration) due to the fact that the coproduced thermal load will be able to be utilized throughout the year, thus increasing the operation period.

According to Figure 4.7, for a typical ICE system and assuming that nowadays gas prices fall within the region of 0.05–0.08 EUR/kWh, combined generation investments will be feasible if electricity prices are over 0.11–0.14 EUR/kWh assuming full heating mode, or over 0.13–0.18 EUR/kWh with cooling mode. For low electricity prices (<0.06 EUR/kWh), cooling mode can be profitable even when heating mode is not, due to the fact that a higher capacity factor, that is, a higher coverage of the loads by the cogeneration equipment, is assumed. In other words, the added value of CHP is not based on the inherent increased efficiency—after all, conventional low-temperature

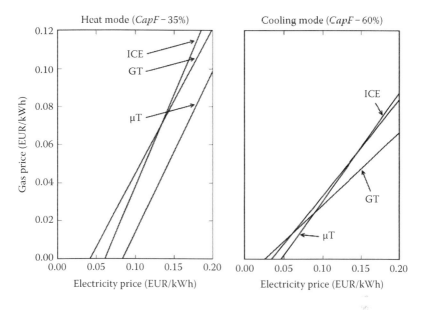

FIGURE 4.7
Investment break-even point for different prime movers and operation modes. (Adapted from Kavvadias, K.C., *Energy*, 115(3), 1632, 2016.)

TABLE 4.3

Cost Assumptions for Technology Comparison

Parameters of Equation	Variable	ICE	GT	µT
Capital cost of prime mover	$C_{eq\,CHP}$ (EUR/kWe)	1600	1100	2200
Capital cost of absorption chiller	$C_{eq\,ab}$ (EUR/kWc)	400	400	400
Overall efficiency	η_{CHP}	90%	82%	85%
Heat-to-power ratio	HPR	1.2	2	0.7
Capacity factor	CapF (%)	Depends on load (assumed 35% for heat only and 60% for heat and cooling)		
Discount rate	i (%)	10		
Investment lifetime	n (years)	20		

Source: Adapted from Kavvadias, K.C., *Energy*, 115(3), 1632, 2016.

heat-driven absorption chillers have very low efficiency—but on the value that the dispatch flexibility adds to the system.

As an example, the evolution of electricity and gas prices (where applicable) from countries reporting to Eurostat is presented in Figure 4.8 and the derived price ratios in Figure 4.9 As a reference, the operational feasibility

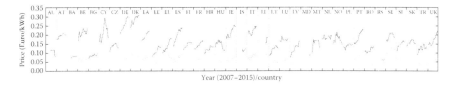

FIGURE 4.8
Time series of natural gas prices (top dark line) and electricity prices (bottom light line).
(The prices shown belong to the high consumption band of the nonindustrial pricelist, which
refers to autoproducers of the tertiary sector: heat > 200 GJ and electricity > 15,000 kWh.)

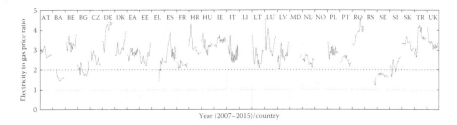

FIGURE 4.9
Historical data of PriceRatio. The bottom and top vertical dotted lines show the operational
viability limit of a typical ICE-based CHP unit based on results of Equation 4.10 for full heating
or cooling mode, respectively.

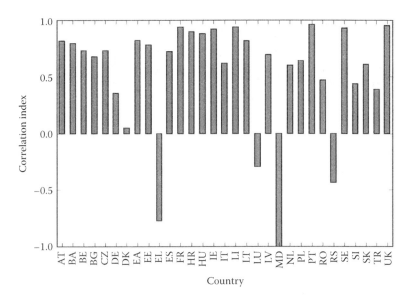

FIGURE 4.10
Correlation of historical data (2007–2015) for EU-28 electricity and gas prices.

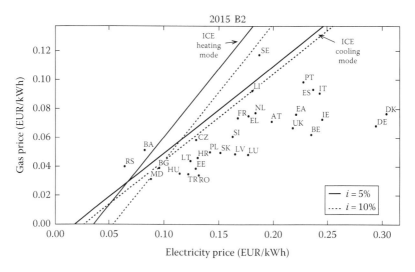

FIGURE 4.11

Gas and electricity prices for 2015. The black lines represent the investment break-even points based on Equation 4.13 with or without production of cooling for two discount rates (5% and 10%).

limits are shown for full heat mode or full cooling mode as presented in Figure 4.3. Currently, it seems that there are a few countries that are close to the operational feasibility point. Indeed, countries such as Bulgaria, France, and Sweden have low market share of CHP autoproducers due to low price ratios. In most cases, the fluctuation of electricity and gas prices has a positive correlation (Figure 4.10). In some cases, there is a weaker correlation due to either an inaccurate pricing mechanism of natural gas or a smaller dependence of electricity production from fossil fuels.

Similarly, the investment driving force for CHP autoproduction is shown in Figure 4.11 for all countries. For the sake of clarity, only one prime mover technology is shown for two discount rates and two operation modes. The further the point from the line, the less attractive an investment is. Countries like Denmark and Germany have the strongest driving force for investments in autoproducer CHP technologies. A group of countries that is close to the break-even line may not have a strong driving force that can justify the risk of future investments without effective policies.

4.2.2 Renewable Distributed Generation

Renewable energy sources (RES) can act as DG. The most popular RES are rooftop photovoltaic (PV) systems due to their ease of installation, small and discrete size, and modular type. Wind energy installations and mini hydro plants are not so widespread for commercial and residential consumers as they have a bigger visual impact and a more complicated installation. For this reason, only the competitiveness of distributed PV is examined in the following section.

4.2.2.1 Competitiveness in Photovoltaics

As in most of the renewable energy technologies and contrary to conventional fossil fuel technologies, fuel costs are nonexistent. As a result, the efficiency* does not directly affect the competitiveness of the energy generation. For an installation of given capacity, an efficiency of 1% will be as economic as an efficiency of 100% as the fuel has zero cost. Although economics are not directly affected, other aspects should be taken into account; for example, for a photovoltaic panel with double efficiency, you will need half the footprint, which may consequently reduce the capital costs. The most critical variables for competitiveness, as in all renewable energy sources, are usually the capital costs ($/kWp) and an operational performance metric, which shows the availability of the resource throughout the year. This is captured by the capacity factor. As a general rule of thumb, the higher the latitude (the further away we move from the equator), the less is the irradiation and consequently the smaller the capacity factor. Irradiation maps are available for a more accurate estimation of production (Huld et al. 2012).

In order to assess the comparative competitiveness of solar energy compared to the electricity provided by the grid, the concept of grid parity is introduced (Yang 2010). According to this, the grid parity price is defined as the point where the total cost to consumers of one technology is equal to the retail grid electricity price (Breyer and Gerlach 2012; Yang 2010). It is believed that consumers will start transitioning to cleaner energy technologies if they can get cheaper electricity.

Photovoltaic installations have seen a very big decrease in costs during the last decades. This technology has the highest learning rate of all energy technologies. For the last 35 years, each time the panel price has dropped by 21%, the worldwide panel production has doubled (Pillai 2015). One must also take into account the big growth rate of the PV industry, which has caused a price drop of more than 45% in the last 15 years. The performance is also getting better in terms of improvements in lifetime, better high-temperature or low-light performance, etc.

In Figures 4.12 through 4.14, the dots show the prices for different countries as reported by Eurostat for the second semester of 2015 for residential, commercial, and industrial consumers, respectively, versus the average capacity factor for each country. The curves represent the levelized cost of

---

* Efficiency here is defined as the ratio of produced energy to primary energy input, for example, in the case of photovoltaics, electricity produced divided by the solar irradiation.

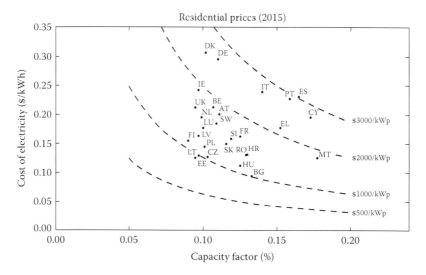

FIGURE 4.12
Electricity prices of residential sector for various countries in 2014 related to the photovoltaic generation effectiveness. The black lines represent the investment break-even points for different capital costs.

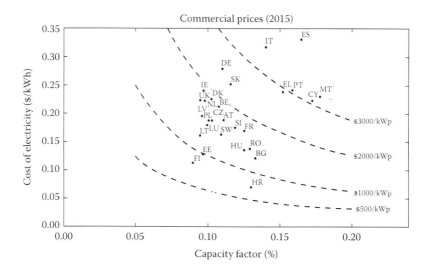

FIGURE 4.13
Electricity prices of commercial sector for various countries in 2014 related to the photovoltaic generation effectiveness. The black lines represent the investment break-even points for different capital costs.

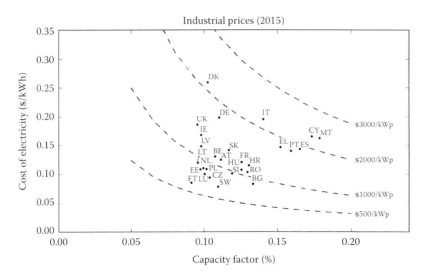

FIGURE 4.14
Electricity prices of industrial sector for various countries in 2014 related to the photovoltaic generation effectiveness. The black lines represent the investment break-even points for different capital costs.

electricity produced by PV for different capital costs.* If a country is above a line, PV can provide electricity at a better price than the grid.

Nowadays, the cost of turnkey solutions varies between 1000 and 2000 EUR/kWp, so many countries have already passed the grid parity point. As with all capital-intensive investments, PV benefits from low discount rates since it lowers its levelized costs. If the equipment costs continue to fall and grid electricity price continues to inflate, even industrial consumers, who usually benefit from attractive rates, will be able to self-consume at a lower cost than the grid. In that case, if self-consumption is increased, for example, by shifting the demand of using electrical appliances like washing machines during the peak sun hours, the economics of the investment will be much more attractive.

Since the middle of the previous decade, feed-in tariffs have been the main policy instrument that has driven the explosive development of PV. Consumers would benefit from selling electricity to the grid at a much increased price, which could even be ten times more than the marginal price of the electricity system. Nowadays, there is a tendency to cut feed-in tariffs in favor of a new market structure where photovoltaic is used for *self-consumption*. According to this, consumers are urged to use their own energy as long as it is more attractive than the grid electricity (Masson et al. 2016).

---

* Levelized costs were estimated using the BNEF small-scale PV and PV with storage (SSPVS) economic model with the following parameters: WACC 10%, lifetime 25 years, self-consumption 45%, PV OPEX 0.01 EUR/kW/annum.

4.3 Overview of Policies, Concerns, and Recommendations

4.3.1 Status and Barriers

DG is more prevalent in the industrial and tertiary sector and to a smaller extent in the residential sector.

Figure 4.15 shows the share of electricity generation from CHP technologies for 2012 sorted by the CHP autoproduction share. The evolution of the installed capacity of CHP technologies, along with the CHP share of different commercial consumers for 2014, is shown in Figure 4.16. The dominance

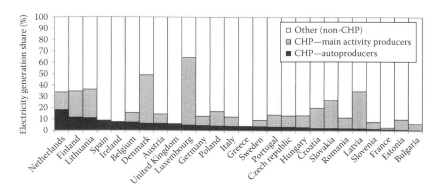

FIGURE 4.15
Share of electricity generation of CHP autoproducers and main producers. (Data from Eurostat, Combined heat and power data, 2014, http://ec.europa.eu/eurostat/statistics-explained/index.php/Electricity_and_heat_statistics. Accessed on August 2015; Adapted from Kavvadias, K.C., *Energy*, 2016, 115(3), 1632.)

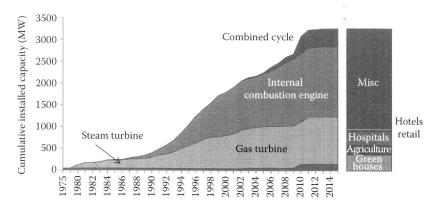

FIGURE 4.16
Installed capacity in 2012 of CHP autoproducers per technology for tertiary sector. (Data from Platts, UDI World electric power plants database, 2015; Adapted from Kavvadias, K.C., *Energy*, 2016, 115(3), 1632.)

of gas-driven technologies in this sector is prevalent. More specifically, the subsectors that are good candidates for such technologies are the so-called MUSH markets (municipalities, universities, schools, and hospitals), as they usually have high demand, high occupancy rate, and suitable heat-to-power ratio. From the sigmoid curve, it may be noticed that the market has passed the phase of exponential growth and has reached its maturity.

The residential sector is not very advanced due to expensive turnkey solutions, longer payback times, and unavailability of sufficient financing mechanisms. Individual houses do not have any good products. It would be economical for small households to have systems of 1–2 kWe but due to economies of scale it is still very expensive; the most affordable devices start from 5 to 6 kW. However, research is very active in this sector and things are expected to change during this decade.

PV is a fast-growing market. The energy produced during 2014 is presented in Figure 4.17 (Eurelectric 2014; EurObserver 2016). The share of the energy produced is much smaller than the share of capacity installed due to the small capacity factor.

There is still potential to be realized if certain barriers are lifted. In general, the barriers of DG technologies fall into one of the following categories (Al-sulaiman et al. 2011; International Energy Agency 2011):

- High initial costs
- Market risks for new technologies
- Imperfect information
- Uncertainty (technical, regulatory, policy, etc.)

According to Baer et al. (2015), the economic challenges of CHP investments are the greatest barriers to viability. Although CHP promises long-term

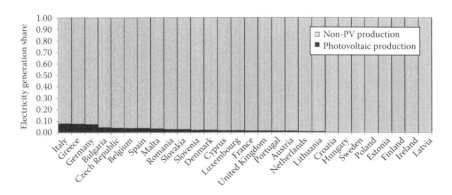

FIGURE 4.17
Share of electricity generation of PV autoproducers for 2014.

energy-bill savings, companies often consider it a greater financial risk because CHP installations have high upfront costs and long payback periods compared to traditional equipment. The recent economic crisis and the difficulties in securing financing have caused companies to become increasingly conservative, with even greater aversion to investments with longer payback periods.

EU Member States have recently reported the barriers of realization of the national potential of high-efficiency cogeneration. The most important barrier—with 17 Member States reporting it—was fuel prices and their volatility (Moya 2013). Other barriers in order of significance are heating demand, law complexity, no support schemes, limited financial resources, regulatory framework uncertainties, etc. A more recent study by Colmenar-Santos et al. (2015) highlights this fact: price volatility and the regulatory framework are the most important barriers and without proper risk mitigation, these projects cannot be easily materialized. Investment opportunities of DG and specifically cogeneration schemes are difficult to evaluate due to the high complexity and multiple sources of risk (Wright et al. 2014). Gas-driven cogeneration operators in the EU have a particular uncertainty because low wholesale electricity prices have coincided with relatively high gas prices, which is causing many plants to operate partially or not at all (Cogeneration Observatory and Dissemination Europe 2015; International Energy Agency 2008).

Despite the benefits mentioned in the previous section, DG may complicate the operation of the distribution system. Large power flows transferred from one place to another may cause congestion in transmission lines. The regulator has to ensure that the design of the system ensures maximum self-consumption, thus minimizing unnecessary infusions to the grid.

4.3.2 Policy Issues and Recommendations for the Expansion of Distributed Generation

According to current data, DG technologies have already been deployed, but further policy support is needed to overcome the current barriers. From a policy perspective, there are a number of actions that can address these problems.

While economics is certainly a fundamental factor, policy also affects the role of DG. On the regulatory side, there is much to be done (Al-sulaiman et al. 2011; Kreith and Goswami 2007).

In order for DG to become an integral part of the energy mix, policymakers must establish clear goals in overall energy policy. There is a need for flexible regulation so that it can facilitate the connection of new DG to the electricity grid. Transparent market conditions are probably the most important policy, which will allow agents to take informed decisions about their investments. This can be complemented with a subsidy reduction in central-ized power plants along with incentives for DG but with a clear phase-out plan. Economic efficiency can be ensured with proper market access and pricing schemes. The main support mechanisms that can be used are the following (Colmenar-Santos et al. 2016):

- *Enhanced feed-in tariffs*: An explicit monetary reward provided for producing electricity from renewables at a rate per kWh some-what higher than the retail electricity rates being paid by the customer.
- *Direct capital subsidies*: Direct financial subsidies aimed at tackling the up-front cost barrier, either for specific equipment or for total installed system cost.
- *Green electricity schemes*: Allow customers to purchase green electric-ity based on renewable energy from the electricity utility, usually at a premium price.
- *Net metering*: The system owner receives retail value for any excess electricity fed into the grid, as recorded by a bidirectional electricity meter and netted over the billing period.
- *Net billing*: The electricity taken from the grid and the electricity fed into the grid are tracked separately, and the electricity fed into the grid is valued at a given price.
- *Sustainable building requirements*: Include requirements on new build-ing developments (residential and commercial) and also in some cases on properties for sale.

The public should be educated on the benefits of DG and special train-ing and jobs around these technologies have to be promoted among all involved stakeholders. A strategy for research innovation and develop-ment of full DG solution should be aligned with national energy goals. This cannot be performed without uniform standards and methodology with common metrics to measure energy savings and environmental benefits.

When trying to integrate DG into big urban areas, special provisions need to be taken into account. Proper measures should also be taken to secure the smooth operation of the grid. This can be in line with the ancillary service market, where consumers can contribute to the maintenance of power qual-ity by providing, for instance, reactive power. In that case, a new way to monetize these services has to be developed. These policies can be in line with general environmental protection and energy efficiency policies in order to achieve the desired long-term goals.

Nomenclature

C	operating costs (EUR)
CC	capital costs (EUR)
CHP	combined heating and power
Co	cooling energy (kW)
COP	coefficient of performance (—)
El	electricity (kW)
F	fuel (kW)
GT	gas turbine
HPR	heat-to-power ratio (—)
ICE	internal combustion engine
PESR	primary energy savings ratio (%)
Th	thermal energy (kW)
α	fraction of cooling from heat (—)
η	efficiency (%)
μT	micro-turbine

References

Ackermann, T., G. Andersson, and L. Söder. 2001. Distributed generation: A definition. *Electric Power Systems Research* 57(3): 195–204.

Al-sulaiman, F.A., F. Hamdullahpur, and I. Dincer. 2011. Trigeneration: A comprehensive review based on prime movers. *International Journal of Energy Research* 35: 233–258.

Badami, M., M. Mura, P. Campanile, and F. Anzioso. 2008. Design and performance evaluation of an innovative small scale combined cycle cogeneration system. *Energy* 33(8): 1264–1276.

Baer, P., M.A. Brown, and G. Kim. February 2015. The job generation impacts of expanding industrial cogeneration. *Ecological Economics* 110: 141–153.

Breyer, C. and A. Gerlach. February 2012. Global overview on grid-parity. *Progress in Photovoltaics: Research and Applications* 21: 121–136.

Capuder, T. and P. Mancarella. July 2014. Techno-economic and environmental modelling and optimization of flexible distributed multi-generation options. *Energy* 71: 516–533.

Cardona, E., A. Piacentino, and F. Cardona. 2006a. Energy saving in airports by trigeneration. Part I: Assessing economic and technical potential. *Applied Thermal Engineering* 26(14–15): 1427–1436.

Cardona, E., A. Piacentino, and F. Cardona. 2006b. Matching economical, energetic and environmental benefits: An analysis for hybrid CHCP-heat pump systems. *Energy Conversion and Management* 47(20): 3530–3542.

Chicco, G. and P. Mancarella. 2009. Distributed multi-generation: A comprehensive view. *Renewable and Sustainable Energy Reviews* 13(3): 535–551.

Cogeneration Observatory and Dissemination Europe. 2015. European Cogeneration Roadmap, http://www.code2-project.eu/wp-content/uploads/CODE-2-European-Cogeneration-Roadmap.pdf. Retrieved from CODE2 website on August 2015.

Colmenar-Santos, A., C. Reino-Rio, D. Borge-Diez, and E. Collado-Fernández. June 2016. Distributed generation: A review of factors that can contribute most to achieve a scenario of DG units embedded in the new distribution networks. *Renewable and Sustainable Energy Reviews* 59: 1130–1148.

Colmenar-Santos, A., E. Rosales-Asensio, D. Borge-Diez, and F. Mur-Pérez. June 2015. Cogeneration and district heating networks: Measures to remove institutional and financial barriers that restrict their joint use in the EU-28. *Energy* 85: 403–414.

EU. 2004. February 2004. Directive 2004/8/EC of the European Parliament and of the Council of 11 February 2004 on the promotion of cogeneration based on a useful heat demand in the internal energy market and amending Directive 92/42/EEC. OJ L 52: 50–60. http://eur-lex.europa.eu/legal-content/EN/ALL/?uri=CELEX%3A32004L0008.

EU. November 2012. Directive 2012/27/EU of the European Parliament and of the Council of 25 October 2012 on energy efficiency, amending Directives 2009/125/EC and 2010/30/EU and repealing Directives 2004/8/EC and 2006/32/EC text with EEA relevance. *Official Journal of the European Union*. OJ L 315: 1–56. http://eur-lex.europa.eu/legal-content/EN/TXT/?uri=celex%3A32012L0027.

Eurelectric. 2014. Electricity consumption EU28 database. http://www.eurelectric.org/factsdb/. Retrieved online on May 2016.

EurObserver. 2016. Photovoltaic barometer. https://www.eurobserv-er.org/photovoltaic-barometer-2016/. Retrieved online on June 2016.

Eurostat. 2014. Combined heat and power data. http://ec.europa.eu/eurostat/statistics-explained/index.php/Electricity_and_heat_statistics, accessed August 2015.

Graves, R., B. Hodge, and L. Chamra. 2008. The spark spread as a measure of economic viability for a combined heating and power application with ideal loading conditions. In Proceedings of ES2008 Energy Sustainability, Paper No. ES2008-54203, pp. 167–171, Jacksonville, FL, August 10–14, 2008.

Huld, T., R. Müller, and A. Gambardella. 2012. A new solar radiation database for estimating PV performance in Europe and Africa. *Solar Energy* 86(6): 1803–1815.

International Energy Agency. 2008. *Combined Heat and Power*. Paris, France.

International Energy Agency. 2011. *Technology Roadmap: Energy Efficient Buildings—Heating and Cooling*. Paris, France.

Kavvadias, K.C. 2016. Energy price spread as a driving force for combined generation investments: A view on Europe. *Energy*, 115(3): 1632–1639.

Kavvadias, K.C., A.P. Tosios, and Z.B. Maroulis. 2010. Design of a combined heating, cooling and power system: Sizing, operation strategy selection and parametric analysis. *Energy Conversion and Management* 51(4): 833–845.

Kreith, F. and D.Y. Goswami (Eds.). 2007. Chapter 2.8 Policies for Distributed Generation. *Handbook of Energy Efficiency and Renewable Energy*. CRC Press, pp. 2–39.

Lončar, D., N. Duić, and Ž. Bogdan. 2009. An analysis of the legal and market framework for the cogeneration sector in Croatia. *Energy* 34(2): 134–143.

Lund, H., A.N. Andersen, P.A. Østergaard, B.V. Mathiesen, and D. Connolly. 2012. From electricity smart grids to smart energy systems—A market operation based approach and understanding. *Energy* 42(1): 96–102.

Lund, H., S. Werner, R. Wiltshire, S. Svendsen, J.E. Thorsen, F. Hvelplund, and B.V. Mathiesen. April 2014. 4th Generation District Heating (4GDH). *Energy* 68: 1–11.

Mancarella, P. February 2014. MES (Multi-Energy Systems): An overview of concepts and evaluation models. *Energy* 65: 1–17.

Masson, G., J.I. Briano, and M.J. Baez. 2016. Review and analysis of PV self-consumption policies. Report IEA-PVPS T1-28:2016.

Moya, J.A. September 2013. Impact of support schemes and barriers in Europe on the evolution of cogeneration. *Energy Policy* 60: 345–355.

Palzer, A., G. Westner, and R. Madlener. August 2013. Evaluation of different hedging strategies for commodity price risks of industrial cogeneration plants. *Energy Policy* 59: 143–160.

Pepermans, G., J. Driesen, D. Haeseldonckx, R. Belmans, and W. D'haeseleer. 2005. Distributed generation: Definition, benefits and issues. *Energy Policy* 33(6): 787–798.

Piacentino, A. and F. Cardona. 2008. An original multi-objective criterion for the design of small-scale polygeneration systems based on realistic operating conditions. *Applied Thermal Engineering* 28(17): 2391–2404.

Pillai, U. July 2015. Drivers of cost reduction in solar photovoltaics. *Energy Economics* 50: 286–293.

Platts. 2015. UDI World electric power plants database. http://www.platts.com/products/world-electric-power-plants-database.

Sezgen, O., C.A. Goldman, and P. Krishnarao. 2007. Option value of electricity demand response. *Energy* 32(2): 108–119.

Smith, A.D., N. Fumo, and P.J. Mago. 2011a. Spark spread—A screening parameter for combined heating and power systems. *Applied Energy* 88(5): 1494–1499.

Smith, A.D., P.J. Mago, and N. Fumo. 2011b. Emissions spark spread and primary energy spark spread—Environmental and energy screening parameters for combined heating and power systems. *Applied Energy* 88(11): 3891–3897.

US Midwest CHP Application Center. 2007. Combined heat and power resource guide for hospital applications. https://energy.gov/eere/amo/downloads/combined-heat-and-power-chp-resource-guide-hospital-applications-2007. Retrieved online on June 2015.

Wright, D.G., P.K. Dey, and J. Brammer. July 2014. A barrier and techno-economic analysis of small-scale BCHP (Biomass Combined Heat and Power) schemes in the UK. *Energy* 71: 332–345.

Wu, D.W. and R.Z. Wang. 2006. Combined cooling, heating and power: A review. *Progress in Energy and Combustion Science* 32(5–6): 459–495.

Yang, C.-J. 2010. Reconsidering solar grid parity. *Energy Policy* 38(7): 3270–3273.

Yao, R. and K. Steemers. 2005. A method of formulating energy load profile for domestic buildings in the UK. *Energy and Buildings* 37(6): 663–671.

5

Are Smart Grids the Holy Grail of Future Grid Mix? Economic, Environmental, and Regulatory Opportunities for Smart Grid Development in Northwestern Europe

Michele Moretti, Nele Witters, Steven Van Passel, and
Sylvestre Njakou Djomo

CONTENTS

5.1 State of the Art and Driving Forces for a Smarter Energy System in Europe

In light of the European Union's objectives on sustainable energy production and supply, several member states have been advocating for the integration of renewable generation, distributed generation, and smart grid (SG) technologies. SGs are recognized by scientists, nongovernmental organizations, and policy makers as a pivotal tool for driving Europe toward an energy-efficient and low-carbon economy. To accelerate the development of SG benefits in Northwest Europe (NWE) and to encourage communities and small-medium enterprises (SMEs), there is a need for a deeper understanding of (1) the level of SG development, (2) the economic and environmental benefits society can receive from SG implementation, and (3) the policies put in place to support SG integration. Three main critical points were recognized in the European SG integration proposition: (1) the lack of overall applicable economic and environmental data, (2) the large amount of components required to fully integrate SG into the existing grids infrastructure, and (3) the existence of several regulatory barriers that limit SG development in NWE. This chapter provides a summary of the potential benefits arising from the development of SG technologies in NWE. More specifically, it consists of three parts: a review of existing economic and environmental impact assessments of SGs, a life cycle assessment (LCA) of the deployment of smarter grid technologies in Belgium, and an analysis of the regulatory status quo and the existing barriers for the deployment of SGs in Northwestern European countries.

5.2 Economic and Environmental Benefits of Adopting SG Technologies*

5.2.1 Literature Review of Existing Impact Assessments

SGs play a central role in the process of transforming the functionality of the current electricity supply systems. The main advantages of SG can be

_____
* This section was partially derived from Moretti et al. (2016).

summarized as follows: improved reliability and security, shift in peak load, enhanced efficiency, and decreased greenhouse gas (GHG) intensity of electricity production and supply systems. However, implementing SG requires the deployment of many new, costly, energy-consuming technologies and devices. The increasing number of studies and reports on SG present widely varying estimates of costs, benefits, and environmental impacts, which leave policy makers struggling with conflicting advice about SG deployment. The chapter is based on recent study by Moretti et al. (2016) and seek to summarize published works on the economic and environmental impacts of SG.

Hence, trying to elucidate this issue, this paragraph analyzes published research (from 2000 to 2014) on the economic and environmental impacts of SG by looking at the different methodologies used. This way, we are able to (1) identify the gaps in SG impact assessment research and (2) understand the origin of variation in the estimates of costs and GHG savings across studies. With these goals, this chapter provides a general overview of SG contribution to the improvement of energy efficiency and environmental impact reductions. When assessing the economic and environmental impacts of SG, tools such as LCA or cost–benefit analysis (CBA) are most used. Consequently, there is a need to have a clear definition of a smart grid as a starting point for conducting these quantitative analyses.

The European Technology Platform Smart Grid (smartgrid.eu, 2015) defines an SG as "an electricity network that can intelligently integrate the actions of all users connected to it—generators, consumers and those that do both—in order to efficiently deliver sustainable, economic and secure electricity supplies." This technologically advanced network facilitates the massive integration of electricity generation (e.g., PV and wind), and application (advanced metering infrastructure, electric vehicles, and heat pumps) technologies, to the current electricity grid (Wang et al., 2011). The current European electricity grid is a traditional radial energy flow, characterized by four main links: generation, transmission, distribution, and off-take (Personal et al., 2014). The aim of these infrastructures is to produce electricity reliably and at a reasonable cost (Blumsack and Fernandez, 2012), but these infrastructures are inadequate to meet the needs established by the European energy policy, namely, (1) achieving highest levels of reliability, (2) increasing energy efficiency, (3) increasing the share of renewable energy, (4) empowering consumers, and (5) building a European integrated energy market (Basso et al., 2013; Blumsack and Fernandez, 2012; Cunjiang et al., 2012; European Commission, 2010; Fouquet and Johansson, 2008). To face the infrastructure inadequacy problem, the "Smart Grid Task Force" was created in 2009 and the new energy vision for a more resilient Energy Union (European Commission, 2015a) was established in 2015 by the European Commission with the aim of implementing SG and related regulations all over Europe. In compliance with these efforts, several dispositions are also taken to favor decentralized generation systems and advanced metering infrastructures (i.e., smart meters). These technological innovations are considered key issues for increasing the share of renewable energy, improving energy efficiency of the grids, and thus reducing the environmental impacts and economic costs of power generation and supply (Faruqui et al., 2011; Hidayatullah, 2011; Simoes et al., 2012). Despite the earlier mentioned benefits, initiatives and investments for the transaction to a smarter energy system across the EU have only started in the last decades (Faruqui et al., 2010; Simoes et al., 2012). As reviewed by Sun et al. (2011), several countries and companies provided guidance for a comprehensive assessment of SG impacts. Moreover, understanding the balance between costs and benefits (economic, environmental, and social) as well as addressing the regulatory challenges is imperative for decision-making at regional and national levels. To clarify and understand the magnitude and variability of economic and environmental impacts and to underline the parameters that drive the results of SG impact assessment, this paragraph provides (1) an examination of the methodologies used to assess the cost and benefits and the environmental impacts of SG, (2) an analysis of the gaps in SG impact assessment research, and (3) recommendations for future research.

Of the 177 potential articles and reports that were initially identified, only those that (i) were written in English, (ii) contained quantitative estimates of economic costs or environmental impacts, (iii) presented the methodology to derive the economic costs or environmental impacts were retained. Twelve papers meeting these criteria were then subjected to further analysis and data extraction. Data relating to the methodologies used, the segments of the electricity network (generation, transmission, distribution), and the technological devices, as well as the definition of SG, were extracted and entered into an Excel spreadsheet. Moreover, the reported data on energy savings, GHG emissions, economic cost, and benefits were elicited in order to obtain comparable results among the different studies (see Table 5.1).

5.2.2 Critical Evaluation of the Outcomes of the Economic and Environmental Impact Assessment of SGs

The economic and/or environmental outcomes of selected studies have been extracted and classified, according to their availability, in different categories: economic costs and benefits (M€/year), energy saved (MJ/kWh), GHG emissions (gCO_2/kWh), and other air emissions (e.g., PM, NO_x, SO_x, and g/kWh).

Several studies define SG by reporting its principal characteristics: (1) optimizing power supply and delivery, (2) automatically minimizing losses through transmission and distribution, (3) providing instantaneous damage control, and (4) accommodating new off-grid alternative energy

TABLE 5.1

List and Main Characteristics of the Examined Publications (N = 12)

Type of Analysis	Methodology	SG Definition	System Boundary	Technology Included	Assumptions	Country/Region	Reference
Economic and environmental	CBA/OP		Grid to consumers for demand-response participation	RET	Three DR programs (demand bidding, ancillary services, and DRSP) combined with wind generation. 41-bus radial system with 1 substation feeding a rural area (peak load 16.8 MW). The system includes 1 sub-station (peak load 16.8 MW), 7 wind power plants (power rated 1.1 MW), and 2 diesel generators (power factor 1–0.9).	Canada	Zakariazadeh et al. (2014)

(Continued)

TABLE 5.1 (*Continued*)

List and Main Characteristics of the Examined Publications (N = 12)

Type of Analysis	Methodology	SG Definition	System Boundary	Technology Included	Assumptions	Country/ Region	Reference
Economic and environmental	CBA	"Smart Grid" refers to a modernization of the electricity delivery system so that it monitors, protects, and automatically optimizes the operation of its interconnected elements—from the central and distributed generator through the high-voltage transmission network and the distribution system, to industrial users and building automation systems, to energy storage installations, and to end-use consumers and their thermostats, electric vehicles, appliances, and other household devices.	How SG technologies in the distribution affect the generation expansion	FFT, RET, ICT	Different combinations according to the level of penetration of SG devices into the current grid and according to different possible nondominated functions of smart technologies used (Pareto set).	United States	Tekiner-Mogulkoc et al. (2012)

(*Continued*)

TABLE 5.1 (*Continued*)

List and Main Characteristics of the Examined Publications (N = 12)

Type of Analysis	Methodology	SG Definition	System Boundary	Technology Included	Assumptions	Country/Region	Reference
Economic	CBA		Whole Power Industry	RET	Shift-load for pick to off-pick for Czech household.	Czech	Adamec et al. (2011)
Economic	CBA		Whole Power Industry	ICT (AMI)	(a) AMI costs: mean value €120–450 from household and non-household meters. (b) Demand reduction due to dynamic pricing form 8%–10% to 60%–90%. (c) Avoided cost of capacity 87 euro/kW year.	EU	Faruqui et al. (2010)
Economic and environmental	CBA		Yokohama-wide energy system	RET, ICT	(a) 6% of energy-use reduction. (b) All consumers will change their behavior. (c) Energy-use reduction of 6%. (d) Electricity price $0.21/kWh. (e) Energy savings of $100 per barrel.	Japan	Farzaneh et al. (2014)

(*Continued*)

TABLE 5.1 (*Continued*)

List and Main Characteristics of the Examined Publications (N = 12)

Type of Analysis	Methodology	SG Definition	System Boundary	Technology Included	Assumptions	Country/Region	Reference
Environmental	Estimation based on energy savings forecasting		TSO and DSO	RET, EV	(a) Energy savings from 9 to 150 kWh and CO_2 emissions based on country statistical data. (b) 5% of the power loss during transmission and distribution (total 9.72%) will be saved using distributed generation. (c) Energy savings and GHG emissions are estimated considering an average consumption of 3000 kWh/household year.	Hungary (EU)	Görbe et al. (2012)

(*Continued*)

TABLE 5.1 (*Continued*)

List and Main Characteristics of the Examined Publications (N = 12)

Type of Analysis	Methodology	SG Definition	System Boundary	Technology Included	Assumptions	Country/ Region	Reference
Environmental and economic	LCA and eco-cost estimation		Home electricity management systems (HEMS) production, use, and disposal considering and average Dutch household consumption patterns	ICT	Three HEMS systems (energy monitoring, multifunctional HEMS, energy management device). (a) Router, PC, and smart meter were not included in the system boundaries. (b) The economic profit is calculated as a 10% energy savings. (c) Yearly energy consumption: 3,500 kWh and 52,800 m^3 natural gas. (d) Increase in energy consumption: 1.5% per year.	Netherlands (EU)	Van Dam et al. (2013)

(*Continued*)

TABLE 5.1 (*Continued*)

List and Main Characteristics of the Examined Publications (N = 12)

Type of Analysis	Methodology	SG Definition	System Boundary	Technology Included	Assumptions	Country/Region	Reference
Economic	CBA		Generation, transmission system operator (TSO), distribution system operator (DSO) and consumption	ICT, EV	(a) One million consumers involved. (b) AMI is phased in gradually over a 5-year time horizon. (c) Costs include direct smart meter operational benefits and consumer-driven benefits according to the mix of technologies they use.	United States	Faruqui et al. (2011)
Economic	Transaction cost			EV	(a) 16 kWh vehicle battery pack. (b) Perfect market information: The value includes the degradation costs of the battery pack ($4.2/kWh). Battery replacement $5000. (c) Electricity price for $140–$250.	United States	Peterson et al. (2010)

(*Continued*)

TABLE 5.1 (*Continued*)

List and Main Characteristics of the Examined Publications (N = 12)

Type of Analysis	Methodology	SG Definition	System Boundary	Technology Included	Assumptions	Country/Region	Reference
Economic	CBA	"Smart Grid" refers to a modernization of the electricity delivery system so that it monitors, protects, and automatically optimizes the operation of its interconnected elements—from the central and distributed generator through the high-voltage transmission network and the distribution system, to industrial users and building automation systems, to energy storage installations, and to end-use consumers and their thermostats, electric vehicles, appliances, and other household devices.	Fully operational SG	ICT, RET	The costs include the infrastructure to integrate distributed energy resources (DER) and to achieve full customer connectivity but exclude the cost of generation, the cost of transmission expansion to add renewables and to meet load growth, and a category of customer costs for smart-grid-ready appliances and devices. (a) The deployment of new technologies is considered a steady process at 2010. (b) The decreasing prices of new technologies have been estimated. (c) Maintenance costs have been included. (d) NPV for benefits estimated based on 2010 price level.	United States	Gellings (2011)

(Continued)

TABLE 5.1 (*Continued*)

List and Main Characteristics of the Examined Publications (N = 12)

Type of Analysis	Methodology	SG Definition	System Boundary	Technology Included	Assumptions	Country/Region	Reference
Environmental	Estimations based on ICT market penetration	A unified communications and control system on the existing power delivery infrastructure to provide the right information to the right entity (e.g., end-use devices, T&D system controls, customers, etc.) at the right time to take the right action. It is a system that optimizes power supply and delivery, minimizes losses, is self-healing, and enables next-generation energy efficiency and demand-response applications.	The whole energy sector	RET, ICT	ITC device penetration level: (a) 25%–75% of devices for direct communication with consumers. (b) 25%–50% of devices for reducing line losses. (c) 5%–25% of devices for continuous maintenance of commercial building equipment. (d) Potential peak demand reduction from 5% to 20%.	United States	EPRI (2008)

(*Continued*)

TABLE 5.1 (*Continued*)

List and Main Characteristics of the Examined Publications (N = 12)

Type of Analysis	Methodology	SG Definition	System Boundary	Technology Included	Assumptions	Country/Region	Reference
Environmental	Estimation on historical data and future scenario		The whole power sector (virtual power plant[a])	RET	Electricity savings: (a) In lighting: 55%. (b) In motor system efficiency: 30%. (c) In electric appliances: 10%. (d) In line losses: 10%–20%. (e) Load management: 1%.	China	Yuan and Hu (2011)

FFT, fossil fuel technology (coal, natural gas, oil); NP, nuclear power; RET, renewable energy technology (solar, wind, biomass, hydropower); ICT, information and communication technology (router, sensor, smart meter) Application technology (e.g. electric vehicle); NA, network assets.

[a] Defined as a set of devices or equipment that allow users to.

sources (Amin and Stringer, 2008; Hledik, 2009; Tekiner-Mogulkoc et al., 2012). Only EPRI (2008) and Gellings (2011) provide a comprehensive definition, describing SG as "a unified communications and control system on the existing power delivery infrastructure to provide the right information to the right entity (e.g. end-use devices, transmission and distribution system controls, customers, etc.) at the right time to take the right action" (EPRI, 2008). All these definitions are fully complementary, but for scientific communication purposes, a more detailed definition of SG could accomplish the methodological needs for assessing the economic and environmental impacts of SG, especially in defining the boundaries of the system. The majority of studies focus on economic impacts (50%), followed by studies on environmental impacts (33%), while the remaining studies (17%) addressed both the economic and environmental impacts. The focus on economic and environmental impacts reflects the greatest concern of private/public stakeholders and regulators about the cost-effectiveness of SG investments (Livieratos et al., 2013; Sullivan and Schellenberg, 2011). Both in the United States and Europe, investments are needed to roll out a fully operating SG. Langheim et al. (2014) reported an estimated investment of $300–$500 billion for the modernization of the U.S. grid, while Faruqui et al. (2010) estimated a gap of around 10–15 billion euros between costs and benefits for achieving the full smart meter penetration in Europe by 2020. Studies adopting the CBA (n = 5), except Farzaneh et al. (2014), assumed a time frame of 20 years, while those using other methods (e.g., transaction cost) chose a time frame according to data availability. The type of technologies included (electric vehicles, renewable energy sources, ICT devices, etc.) and the segments of the grid considered (generation, transmission, distribution, and end use) are other sources of variability in the estimates of costs or environmental impacts between the analyzed studies.

The most considered system boundaries are the whole country electricity generation and supply mix, but few studies refer to smaller spatial levels such as single household consumptions or grid-to-consumer distribution systems. Five out of twelve studies focus on GHG emission reduction, while three cover other air pollutants in addition to the GHG emissions.

Figure 5.1 compares the GHG emissions of the baseline and the forecasted scenario. The SG development will produce an expected reduction ranging from 10 to 180 gCO_2/kWh with a median value of 89 gCO_2/kWh (Figure 5.1). GHG emissions estimates across the studies, differ according to assumption about yearly household consumption, proportion of power generated by the different sources (coal, gas, wind, solar panels, etc.) (EPRI, 2008; Görbe et al., 2012; Hledik, 2009; Tekiner-Mogulkoc et al., 2012; Van Dam et al., 2013), types of ICT devices used, penetration in the current grid (Görbe et al., 2012; Hledik, 2009; Tekiner-Mogulkoc et al., 2012), and uncertainty related to consumer participation in demand-response (DR) and electricity generation technologies.

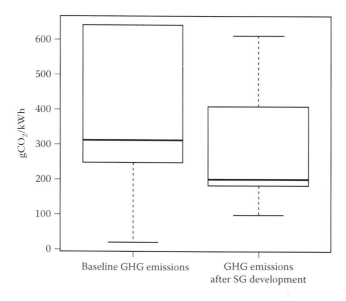

FIGURE 5.1
Distribution of GHG emission outcomes for the studied and baseline scenarios (n = 5).

According to Farzaneh et al. (2014), the more relevant reductions occur for NO_x and SO_2 with a decrease, respectively, of around 4 and 10 kt/year. Tekiner-Mogulkoc et al. (2012) found less significant benefits in terms of NO_x and SO_2 emission reductions by shifting energy demand, respectively, with 5% and 10% from peak load to off-peak load. Both studies estimated a reduction in NO_x ranging on average from 0.6 to 0.9 kt NO_x/year, while the SO_2 emission reductions equal on average 2 kt SO_x/year for both demand shift scenarios. Energy savings have been addressed by Van Dam et al. (2013) who performed an LCA of different ICTs for consumers' participation in a DR system. They reported that an average of 0.26 MJ/kWh electricity can be saved by integrating ICT devices to home energy systems.

Considering the same level of energy consumption reduction during peak load for the whole power system (generation, transmission, and distribution) for the city of Yokohama (Japan) and the State of New Jersey (United States), Farzaneh et al. (2014) and Tekiner-Mogulkoc et al. (2012) estimated roughly the same outcomes in terms of GHG emissions, respectively, 200 and 196 gCO_2/kWh.

Implementing SGs can play an important role in reducing CO_2 emissions, non-CO_2 greenhouse gases (e.g. CH_4 emissions), and other air pollutants such as SOx and NOx. In this regard, Pratt et al. (2010) predicted a reduction in electric utility electricity and CO_2 emissions by 2030 in the United States, attributable to direct energy consumption and CO_2 emissions reduction (12%) and indirect impacts reduction (6%) due to SG. Jiahai and Zhaoguang (2011) studied the low carbon electricity development based on SG construction in China. They found that electricity consumption would decrease from 1,279 to 1,054 KWh per 10,000 yuan GDP, whereas the CO_2 intensity of electricity production in China would reduce from 0.70 to 0.52 tons per MWh during the 2010–2030 periods, as a result of efficiency enhancement and energy structure improvement. According to the International Energy Agency, considering direct and indirect emission reductions, SGs provide the possibility to gain net annual emission reductions of 0.7–2.1 Gt per year of CO_2 by 2050 (International Energy Agency, 2013).

The economic assessment of SG usually refers to the whole power system at the country or smaller spatial level. The economic costs and benefits of SG development in the EU and the United States were addressed by Gellings (2011) and Faruqui et al. (2009). Adamec et al. (2011) and Tekiner-Mogulkoc et al. (2012) targeted their studies to smaller power generation systems by addressing the economic profitability of SG development in the Czech Republic and New Jersey. Farzaneh et al. (2014), by accounting for the forecasted penetration level of advanced metering infrastructure (AMI) and system performances monitoring devices, used CBA to assess the economic and environmental impacts (energy savings, GHG, and other air emissions) of the Yokohama city grid. Zakariazadeh et al. (2014) and Van Dam et al. (2013) focused on the effects of consumer participation to DR systems. The former used a multiobjective optimization function to evaluate the costs of consumer participation to an open electricity market by means of an energy management system that behaves as an aggregator of distributed energy resources. Peterson et al. (2010) used transaction cost analysis to assess consumer participation to energy storage capability considering the capability of electric vehicle owners to offset their own electricity consumption during high-price periods. Although authors assumed perfect market information (including degradation costs of battery pack and battery replacement costs), they stated that "it appears unlikely that these profits alone will provide sufficient incentive to the vehicle owner to use the battery pack for electricity storage and later off-vehicle use" (Peterson et al., 2010). The comparison of economic costs and benefits derived from the analyzed studies reveals SG as a noneffective solution for an investment. Estimated costs show a maximum variation of ~2 order of magnitude across studies (range 0.01–2.55 million euros/year, median 1.5 million euros/year), whereas estimates of the benefits show a variation of 1 order of magnitude (range 0.01–0.7 million euro/year, median 0.2 million euros/year). Thus, on average, the gap between cost and benefit results equals to 0.6 million euros/year (Figure 5.2). Ten times higher economic costs (average value of 25 million euros) than the estimated average, have been estimated by Zakariazadeh et al. (2014). Nevertheless, the assumptions related to the time scale (varying from 5 to 20 years), the typologies of included costs (Faruqui et al., 2010; Tekiner-Mogulkoc et al., 2012), the share of consumers involved and their responses to DR programs (Adamec et al., 2011; Faruqui et al., 2011), the assumed level of energy saved (Van Dam et al., 2013), the energy-market costs (Farzaneh et al., 2014; Gellings, 2011), the costs,

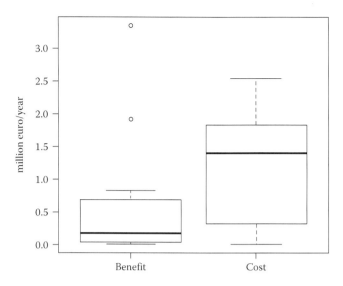

FIGURE 5.2
Distribution of cost and benefit outcomes from the analyzed references (n = 8).

typologies, and replacement costs of ICT devices (Faruqui et al., 2010, 2011) strongly affect the outcomes of the reviewed studies.

5.2.3 Conclusions

SG will provide, among many benefits, significant reductions in CO_2 and other air pollutants. Research and development will be needed to provide suitable modeling techniques/frameworks that are able to provide a measurable environmental benefits of the SG vision. Overall, SGs provide many benefits to customers (Lewis, 2013):

- Savings from lower electricity distribution and retail prices, derived from the efficiencies gained from an SG by distribution and retail companies
- Decreased frequency and duration of power outages
- Capability to become more energy self-sufficient in terms of energy
- Advantages derived from a range of other convenience and comfort-related services like home automation and electric vehicles

All the papers and reports reviewed differ, to some extent, in their approach and the components of the SGs they encompass. Although several studies claim that SGs are not a profitable investment, because the economic costs exceed the benefits (e.g., Adamec et al., 2011; Gellings, 2011), other studies claim that SGs can provide productive users, financial benefits, and employment generation (Faruqui et al., 2010, 2011; Yadoo and Cruickshank, 2012). Considering the key role of SGs in improving efficiency, reliability, and security of energy supply, and due to the major investments needed for SG implementations, the high level of uncertainty about the technological development of ICT devices, and the reliability of distributed generation and customer responses to changes in energy-market price dynamics, there is a need for developing a systematic cost–benefit approach to evaluate the potential benefits of an SG project and justify its application.

To be effective and realistic, the CBA should precisely take into consideration actual data from the available SG pilot projects that have been developed or are currently under development. Some efforts have already been made in this direction (Eurelectric, 2012; European Commission, 2012; Livieratos et al., 2013), but the standardization of the SG impact assessment framework is far from complete. The huge variability in ICTs, the level of penetration of renewable resources (PV, wind, etc.), the regulatory barriers, and the uncertainty in consumer participation to demand-response programs make this process extremely complicated.

5.3 Energy Efficiency and CO_2 Intensity of a Smarter Energy System: Case Study of Belgium

The electricity infrastructure in Belgium and in many EU member states is about 30–40 years old (Battaglini et al., 2009). Moreover, the world demand for energy is expected to increase by 2030, following the population growth, generating a strong increase in energy resources depletion. Additionally, the power sector is one of the major CO_2 emitters, responsible for climate change. Energy efficiency and reliability are therefore critical issues to reduce pressure on the fossil fuel and to decrease the carbon footprint of the power sector.

The current Belgian conventional and smart electricity supply systems were evaluated using life cycle assessment (LCA) framework. LCA is a well-established framework for environmental analysis that aims to quantitatively assess the relationship between a product, process, or activity and the environment. Identifying the product, process, or activity is the starting point of the analysis, which allows defining the input/output flows of materials and energy inside the product system along its whole life cycle, from cradle to grave, from raw material extraction to the product's end of life. This comprehensive "cradle to grave" analysis allows for assessing the contribution of each single production phase to the overall environmental impacts generated by the studied system. The goal of this case study is to quantify

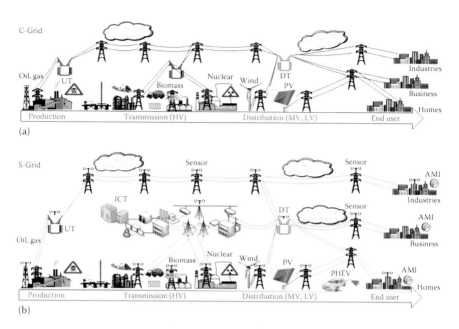

FIGURE 5.3
Schematic representation of the (a) conventional grid (C-Grid) and (b) smart grid (S-Grid).

the physical environmental impacts of an SG system and to compare them with those of the current conventional grid chosen as a reference system. The system boundaries of the two systems are shown in Figure 5.3. The chosen functional unit is 1 kWh$_e$ at the distribution gate. This means that we did not include the impact of electricity use and therefore could not model the effect of consumer behavior on the SG. In fact, there is a high level of uncertainty associated with the home management of ICT devices, the development of distributed energy generation systems, and the consumer participation in the energy system. Environmental impacts were limited to global warming and to the consumption of primary nonrenewable energy. For both the smart and the conventional grid systems, we included all the technologies used for the production, transmission, and distribution of power. The extraction and transport of raw materials for energy production (coal, gas, biomass, uranium, etc.) were not included into the analysis. For the SG system, we included additional ICT devices (sensors, switches, local area networks, and a smart meter) and new infrastructures (e.g., poles and new substations) required for the proper operation of the SG system. Primary data were collected from power producers and from transmission and distribution companies in Belgium (ELIA, ENECO). Secondary data were gathered from the Ecoinvent v2.2 (Frischknecht et al., 2007) and from Hischier et al. (2010). The modeling of environmental impacts was done using Simapro 7.3.3.

5.3.1 Life Cycle Inventory Construction

All the input data related to production, energy transmission, and distribution were collected. Data on ICT devices for the well functioning of the SG system were also gathered. The National Renewable Energy Action Plan of Belgium* (NREAP) expects 13% of penetration of renewable energy by 2020. The level of penetration of ICTs has been defined according to the target forecasted by the NREAP. The collected data include technical data on power generation technologies (including renewable energy), information technology equipment, advanced metering equipment, and transmission cables.

5.3.2 Life Cycle Impact Assessment

The assessed environmental impacts were limited to global warming and use of nonrenewable energy resources. These impacts were evaluated using the IMPACT 2002+ method. Nonrenewable energy consumption is the most commonly used indicator to measure fossil energy used in electricity generation plants, electricity networks, buildings, etc. It is calculated in terms of MJ/kWh_e expressing the amount of nonrenewable energy used to produce a kilowatt-hour of electricity. Typically, global warming potential is reported in the amount of carbon dioxide equivalent (CO_2 eq.) emitted from the production of 1 kWh of electricity. It includes CO_2 and non-CO_2 emissions (e.g., CH_4 and N_2O) related to the production, transmission, and distribution of electric energy. The global warming potential (GWP) is usually calculated over a specific time interval, commonly 20, 100, and 500 years. For this analysis, a GWP over 100 years has been assumed to analyze the climate impact of both conventional and SG systems. Table 5.2 summarizes the data and main assumptions used to model SG development starting from the current Belgian energy grid.

Data on the number of substations in the current Belgian electricity grid were collected from ELIA (Belgian electricity transmission operator). Using data on the number of substations and assuming that six feeders are needed for each substation, the number of new substations needed to handle the growth load was estimated to be 15% of the existing ones. The ones to handle the renewable energy generation growth are inferred to be 1% of the existing ones. Moreover, other assumptions directly linked to the transmission grid have been made as follows: (1) one dynamic thermal rating circuit requires 12 km cable/DTCR, (2) 50% of existing substations will have sensors, and (3) 100% of new substations handling conventional and renewable load growth will have sensors (Table 5.2).

---

* National renewable energy action plan (pursuant to Directive 2009/28/EC). Prepared by the Federal-Regional Energy Consultation Group CONCERE-ENOVER. November 2010. www. buildupeu/pubblications/22818, assessed on the 27th of November 2015.

TABLE 5.2

Main Assumption in the SG Model

Number of Substations and Feeders	Substation Units	Feeders Units	Assumption
Base case (2012)	847	5202	6
New substations to handle load growth (2020)	130.05		15%
New substation to handle renewable energy (2020)	8.67		1%
Feeders			
Base case (2012)		780.3	
New feeders to handle load growth (2020)		52.02	

Number of Dynamic Thermal Circuit Rating (DTCR)

	AC Cables (100–230) Units	Assumption
Base case (2012)	2441	12
Number of DTCR	203.42	

Number of Sensors

	Sensors Units	Assumption
Base case (2012)	433.5	50%
Transmission line sensor for handling renewable energy (2020)	130.05	100%
Transmission line sensor for handling load growth (2020)	8.67	100%

5.3.3 Conventional Grid and SGs

The current Belgian power generation system (hereinafter referred to as conventional grid) is responsible for the emission of almost 281 gCO_2 eq./kWh of produced energy in the atmosphere. The major contributors to this impact are natural gas and coal generation, which account for 65% and 32% of all GHG emissions, respectively. The penetration of renewable energy technologies will decrease the total amount of CO_2 eq. of the Belgian generation grid from 281 to 252 gCO_2 eq./kWh by 10%. This result will be achieved mainly allowing more RES to be connected to the grid and thus reducing the GHG emission intensity of the current grid. In particular, the GHG emitted by both generation systems will decrease with almost 20 and 10 gCO_2 eq./kWh, respectively (Figure 5.4).

The total nonrenewable energy of the conventional grid was 11.5 MJ/kWh, illustrating an overall energy efficiency of grid of about 31%. The low overall efficiency of the Belgian grid can be attributed to nuclear power, which represents about 65% of the generation mix and has an energy efficiency of 46% on a life cycle basis.

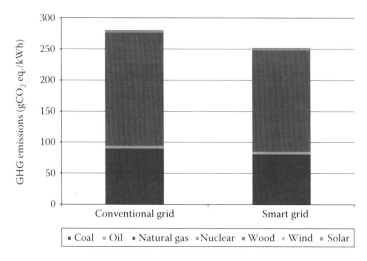

FIGURE 5.4
GHG emissions of the electricity generation system.

The differences between conventional grid and SG are less definite in terms of nonrenewable energy consumption. The conventional grid is responsible for the consumption of almost 11.50 MJ of nonrenewable energy per kWh of electric energy produced. Responsible for this consumption pattern are nuclear and natural gas generation systems, which account for almost 60% and 31%, respectively. The coal generation system shows a marginal role in this impact category (only the 9%) compared to GHG emissions (Figure 5.5).

With the development of SG technologies, the nonrenewable energy consumption of the Belgian power generation system will be reduced to almost 10 MJ/kWh. This reduction will be mainly achieved by increasing the energy efficiency of the nuclear and natural gas generation (Figure 5.5).

The current Belgian transmission network is responsible for the release of almost 5 gCO_2 eq./kWh of transmitted electricity (Figure 5.6). The higher share of GHG emissions occurs in the medium-voltage transmission grid, which accounts for 60% of the whole CO_2 eq. emitted by the conventional transmission network.

According to the SG model used for this study, the GHG emissions of the "smarter" transmission network are higher than the conventional one, accounting for almost 8 gCO_2 eq./kWh (Figure 5.6). The medium-voltage network shows a level of GHG emissions lower than in the conventional grid (1 gCO_2 eq./kWh), while the impact of ICT devices is very high ($\approx$7 gCO_2 eq./kWh), generating the higher impact of the whole "smarter" transmission network. Finally, the overall estimated savings provided by SG penetration into the conventional grid is about 9% for both GHG emissions and nonrenewable energy consumption (base year 2012).

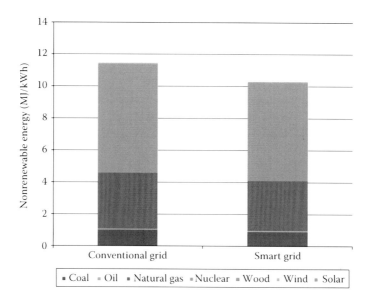

FIGURE 5.5
Nonrenewable energy consumptions of the electricity generation system.

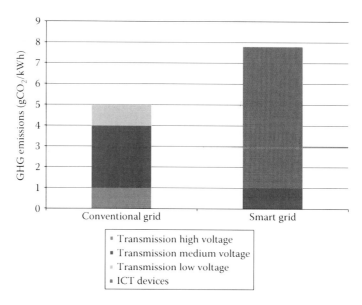

FIGURE 5.6
GHG emissions of the Belgian electricity transmission and distribution system.

5.4 Regulatory Framework and Barriers Regarding SG Applications for Household and SMEs

In light of the European Union's objectives concerning energy efficiency and sustainability, several member states are advocating for and often subsidizing the integration of distributed renewable generation and SG technologies. SG technologies face a number of regulatory barriers that impede their deployment in communities of residential consumers and small and medium enterprises. Therefore, starting from the results of the environmental analysis, the regulatory barriers for further deployment are analyzed. A regional comparison of the regulatory framework of Flanders, the Netherlands, Ireland, and the United Kingdom is conducted, which results in a number of policy recommendations that would facilitate the integration of sustainable technologies in the electricity sector.

First, a regional analysis of the regulatory framework is made by means of the results of a survey, complemented by means of desktop research. This allowed to define the primary regional differences and best practices. Second, policy recommendations and guidelines are formulated based on the results of the environmental analysis and the regulatory framework in place. These general recommendations are applicable in the United Kingdom, Ireland, the Netherlands, and Flanders, as well as in the rest of Europe. The emphasis of the assessment lies on three aspects of regulation that are seen as key for investments and the operation of sustainable technologies by residential consumers and SMEs: (1) the decision to roll out smart metering systems, and its practical implementation, which facilitates the deployment of SG applications within residential communities and SMEs; (2) the structure of the electricity retail price, which provides grid users with direct price signals for the integration of sustainable technologies and the smart control thereof; and (3) the availability of demand–response services, which determines the possibility and compensation for grid users to participate in remunerated services.

The regulatory assessment has uncovered the regional differences across Northwestern Europe and identified major barriers toward further integration of distributed renewable generation and SG technologies. First, most regions have decided upon a smart meter roll-out, but uncertainties following the roll-out strategy impede the further integration of SG technologies. Second, the electricity prices, which are subject to local regulatory provisions, more specifically regarding the network tariff, typically do not adequately incentivize investments in SG technologies. Third, the market entry, where network services in some regions are opening up for new technologies providing flexibility, such as demand-response (DR), does not yet facilitate participation of residential consumers.

5.4.1 Deployment of SG Technologies

In general, no support mechanisms are present concerning the SG hardware that could enable the control of the distributed energy resources, such as heat pump and photovoltaic installation. However, the roll-out of SG infrastructure is closely linked with the roll-out of smart metering equipment, which is closely linked to the regulatory framework and energy-market policies. The EU aims to replace at least 80% of electricity meters with smart meters by 2020 wherever it is cost effective. Therefore, all EU member states were asked to carry out a CBA regarding the roll-out of smart meters by 2012 (Energy Efficiency Directive 2012/27/EC). Table 5.3 provides an overview of the main aspects of the SG roll-out status in each of the regions. Flanders is the only region that did not yet decide upon a roll-out of smart metering systems. The results of the CBAs that were conducted were seen as inconclusive,

TABLE 5.3

Smart Meter Roll-Out Status

Region	Flanders	Ireland	Netherlands	United Kingdom
Smart meter roll-out status	No roll-out plan as the CBA was not conclusively positive. Debate to roll out or not is ongoing.	Ireland aims at a full smart meter deployment by 2019.	Full roll-out of smart metering infrastructure plan over the next 6 years, until 2020	The UK government aims to roll out approximately 53 million smart electricity and gas meters to domestic properties and nondomestic sites in Great Britain by 2020. This will impact approximately 30 million premises.
Deployment strategy	—	Mandatory.	Mandatory (opt-out)	Mandatory.
Metering activity	Regulated.	Regulated.	Regulated	Competitive.
Responsible party for implementation and ownership	DSO.	DSO.	DSO	Supplier.
Responsible for third-party access to metering data	DSO.	DSO.	DSO	Central hub.
CBA outcome	Inconclusive.	Positive.	Positive	Positive.
Financing of the roll-out	—	Network tariffs.	Network tariffs	Funded by suppliers.

Source: European Commission (2014).

and policy makers still debate on the roll-out plan for Flanders. In contrast, all other regions reviewed in this study decided upon a full roll-out toward 2020, after positive CBAs. The Netherlands and Ireland will follow a regulated model, where distribution system operators (DSOs) are responsible for the roll-out, and their operation, and where the costs will be allocated to the consumers by means of the distribution tariff. In contrast, the United Kingdom implements a nonregulated model, putting the roll-out responsibility with the suppliers. Both models have advantages and disadvantages, elaborately discussed in the literature (Reuster et al., 2014).

5.4.2 Electricity Retail Prices

Despite efforts to integrate EU energy markets, household electricity prices vary significantly among the member states. Figure 5.7 illustrates these differences at the end of 2014. Prices range from about 6 Eurocents/kWh in Serbia to more than 30 Eurocents/kWh in Denmark. The average retail price was 21 Eurocents/kWh across the member states. It can be seen that the four regions dealt with in this chapter are on the high end of the scale. Prices are especially high in Ireland (over 25 Eurocents/kWh) and relatively moderate in the Netherlands (approximately 17 Eurocents/kWh).

There are a number of reasons why retail prices vary across EU countries. On the one hand, this is the result of the limited integration of wholesale electricity markets and limited interconnection capacity. This is coupled with the fact that Europe is characterized by strong geographic differences between countries and differences in access to energy sources, which leads to widely varying generation portfolios. Aside from the commodity prices, there are also strong differences in network tariffs. These are largely regulated by national regulatory authorities who, despite some guidelines and laws on the EU level to protect consumers, have quite a lot of freedom in deciding on the tariff methodology. Finally, member states have widely different taxation policies for the consumption and generation of electricity.

At the same time, retail prices in Europe do share a number of characteristics. As a consequence of liberalization and unbundling during the last few decades, there exists a strict divide between commercial and network activities, which are both legal and operational. This division can also be observed in the retail electricity bill, which is typically composed of three major components:

1. Energy commodity component (related to generation and supply of electricity)
2. Network component (related to transmission and distribution)
3. Tax component (including VAT and levies)

Figure 5.8 illustrates the national differences in the shares of commodity and network components. It can be seen that the relative share of network costs

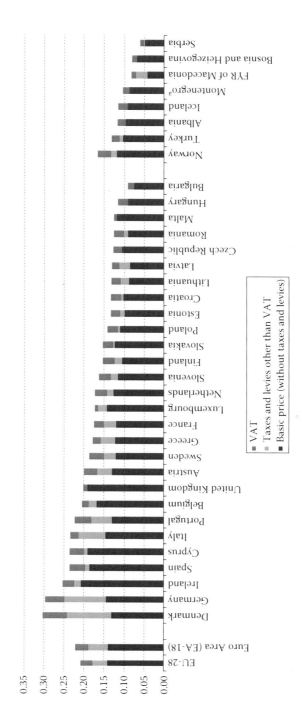

FIGURE 5.7

Electricity retail prices and their structure in Europe. *Note:* Annual consumption: 2500 kWh <consumption < 5000 kWh. [a]Taxes and levies other than VAT are slightly negative and therefore the overall price is marginally lower than that shown by the bar. (From EUROSTAT, 2015.)

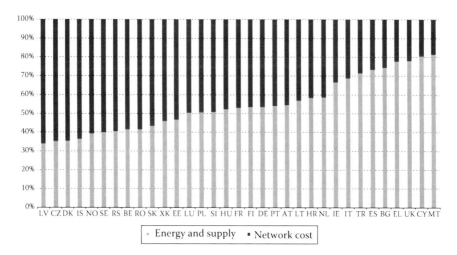

FIGURE 5.8
Relative share of energy and network components in EU retail prices. (From EUROSTAT, 2015.)

in the final price is relatively low in Ireland and the United Kingdom and quite high in Belgium (including Flanders).

In Europe, electricity commodity prices are typically not regulated. Ever since the generation and supply part of the sector has been liberalized and unbundled from network activities, policy makers have relied largely on competitive pressure to keep prices down. However, in many regions, there are some built-in safety mechanisms in the regulatory framework, allowing governments to intervene (e.g., by imposing maximum prices).

The vast majority of electricity suppliers in Europe offer products based on time-of-use (ToU) pricing. These products are usually quite simple and limited to day and night or seasonal rates. More complex ToU schemes are not typically applied to small and residential users. Dynamic pricing (i.e., when the commodity price is subject to changes or indexation during the billing horizon) is also frequently applied. However, the dynamic contracts can be restrained by regulation in terms of updating frequency, in order to protect the consumer against frequent and large price spikes.

With respect to distribution grid tariffs, the allocation of network costs to small and household consumers generally happens based on the following tariff drivers/components:

- Active energy withdrawal (euro/kWh)
- Capacity component (euro/kW) (this includes capacity of the connection and peak usage)
- Fixed component, aimed at recovering fixed costs related to metering and billing

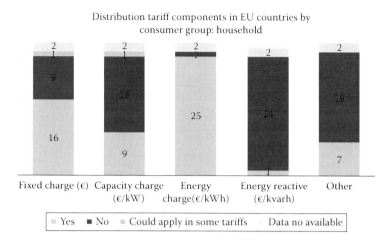

FIGURE 5.9

Household distribution grid tariff components in EU member states. (From European Commission, Study on tariff design for distribution systems, 2015b.)

Several countries have ToU distribution grid tariffs in place, but these are typically not applicable to residential users and are limited to day and night or seasonal rates. Tariffs based on electricity usage during system peak times are not being applied to households in any member state (European Commission, 2015b). Figure 5.9 illustrates the distribution tariff components and the frequency of their usage in household consumer groups in EU member states.

In terms of weight given to the different tariff drivers, there are also large differences across the member states. This ranges from almost purely energy based to almost purely capacity-based tariffs. Note that the Netherlands is the only country in Europe that does not apply any form of energy-based tariff to residential grid tariffs (European Commission, 2015b).

In what follows, the regulatory aspects of commodity prices and (use-of-system) network tariffs for households are discussed in more depth for a number of regions: Flanders, the Netherlands, Ireland, and the United Kingdom. Table 5.4 provides a summary of the main characteristics.

5.4.2.1 Flanders

In Flanders, electricity commodity prices are not regulated, at least not in the classical sense of the word. Suppliers are, by principle, free to set their prices. However, with consumer protection in mind, policy makers have incorporated a set of rules in the electricity market laws (the Electricity Act of April 29, 1999) that limit the ability of suppliers to choose or change price levels. In particular, prices in variable contract types (i.e., when the electricity price is

TABLE 5.4

Regulatory Aspects of Electricity Commodity Prices and Distribution Grid Tariffs in Flanders, the United Kingdom, Ireland, and the Netherlands

	Energy	Distribution (Use of System)
Flanders	• Prices are not regulated, but limitations on dynamic pricing. • Net metering allowed for small users.	• Regulated by Vlaamse Reguleringsinstantie voor de Elektriciteits-en Gasmarkt (VREG) • Energy-based + fixed component • Injection tariffs (>10 kW) • Net metering (<10 kW) • Prosumer tariff (based on generation capacity, <10 kW)
United Kingdom	• Not directly regulated. • Indirect regulations through supply license conditions (obligation to inform consumers of their cheapest rate and potential savings).	• Regulated by Ofgem • Energy-based + fixed component
Ireland	• Not regulated, suppliers can set own tariff regimes. • Suppliers will be obliged to implement a ToU tariff scheme after smart meter roll-out.	• Regulated by Commission for Energy Regulation (CER) • Energy-based + fixed component • ToU tariffs possible (day/night scheme) • Separate urban and rural tariffs
The Netherlands	• Suppliers are allowed to offer dynamic pricing schemes. • The ACM monitors the energy suppliers and energy prices. The energy market is free but the ACM is allowed to set maximum prices if energy pricing is set to high, following the monopolistic character of the market.	• Regulated by ACM • Capacity-based (based on connection capacity) + fixed component • No ToU tariffs • Uniform for residential consumers • No energy/power injection tariff

allowed to change during the course of the contract horizon) are limited, in the sense that price changes (indexations) are allowed maximum four times per year, at the start of each trimester (Art 20).

Regulations, as described in the example earlier, put significant limitations on the ability of commercial players (suppliers, aggregators) to develop active demand products based on dynamic pricing. In principle, it is allowed to charge different prices for different times of the day. One of the simplest forms of this (day and night pricing) is already being applied regularly in Belgium. However, the updating of these prices to real-time market conditions is limited to a maximum of four indexations per year. Similarly, both in fixed and variable contract types, because of maximum prices, commercial players can be limited in their ability to charge cost-reflective prices when market conditions deteriorate.

In Flanders, the regional regulator VREG and the DSOs have shared responsibilities in network tariff setting. The regulator sets the tariff methodology and approves the tariffs calculated and proposed by the DSO. Because of incentive regulation, the tariffs collected by the DSO are limited by a revenue cap, which is reviewed every 4 years. Distribution grid tariffs for households are largely calculated on the basis of active electricity off-take (euro/kWh), with an added fixed charge to cover metering and billing costs. For prosumers, the following rules apply:

- Prosumers with renewable generation capacity below 10 kW are allowed to compensate electricity off-take and injection (net metering). In this case, the electricity meter rolls backward whenever electricity is injected into the grid. For these users, grid tariffs are calculated on the basis of annual net off-take.

- For larger generation facilities, exceeding the 10 kW limit, the prosumer is obliged to install separate meters for injection and off-take. An injection tariff is applicable. At the moment, this tariff is low relative to off-take tariffs (on average 0.5 Eurocent/kWh for injection vs. 0.13 Eurocent/kWh for off-take).

- In order to still receive compensation for electricity injected into the grid, the user has to enter a contract with a supplier who wishes to purchase this electricity. There is, however, no obligation on the part of suppliers to do this.

- From January 2015 onward, an additional grid tariff is applicable for prosumers subject to net metering. A "prosumer tariff," based on the capacity of the PV system inverter, has been introduced in order to increase the contribution of prosumers to carrying network costs. This measure has been quite controversial, as it follows a long period of subsidies and beneficial tariff caution practices toward renewables. In addition, the tariff is often deemed discriminatory toward renewable generators who actively manage their consumption and generation and as such do not contribute as much to network costs.

In general, distribution grid tariffs in Flanders do not encourage the active management of local generation and demand. Instead, they incentivize low-grid off-take (which is not necessarily equivalent to low consumption) and a fit-and-forget approach for local generation. If the investment cost and subsidy mechanisms are attractive enough, local generation capacity is best installed up until the point where annual generation equals annual consumption, such that the meter reading, and by extension the bill, drops to zero. The new prosumer tariff, based on the capacity of the inverter, discourages investment in distributed generation, even where it is beneficial for the grid due to active management of demand.

5.4.2.2 United Kingdom

The methodology for the calculation of distribution grid tariffs is regulated by Ofgem. The regulator determines the allowed revenue and the tariff methodology. A so-called RIIO-ED1 price control is in place, setting the outputs that the 21 DSOs must deliver and the associated revenues they are allowed to collect over an 8-year period. This system replaces the previous RPI-X approach, and the first period started on April 1, 2015, and lasts until March 31, 2023. The objective of this price control model is to ensure sufficient investment in the grid at a reasonable price for the consumer.

In the United Kingdom, distribution grid tariffs for residential consumers are based on active energy off-take and include an additional fixed charge. Connection charges are predominantly shallow (European Commission, 2015b). Electricity commodity prices are not directly regulated. Suppliers are regulated indirectly through their license conditions. In August 2013, Ofgem (the regulator) implemented modifications to the license conditions with the aim of simplifying supplier switching and comparison by residential consumers. Since April 2014, suppliers are obliged to inform consumers of their cheapest rates. This means that there are no (known) regulatory objections to implement alternative pricing schemes.

5.4.2.3 Ireland

The distribution sector in Ireland is characterized by a single, state-owned DSO (ESB Networks Ltd). An incentive-based model is used, based on a 5-year revenue cap. The regulator (CER) is the sole responsible in establishing a tariff methodology and calculating the rates. Network costs are allocated across different system levels and divided by the number of customers in each level. For example, a distinction is made between rural and urban consumers. An urban connection is defined as fed from a three-phase overhead or underground LV network, while a rural connection is characterized by a connection to a single-phase overhead network.

The network tariff structure is predominantly based on energy charges and fixed annual charges (European Commission, 2015b). Capacity-based charges only apply to larger consumers (larger businesses, car manufacturers, etc.). ToU tariffs do exist but are limited to day and night schemes.

No information is found on regulatory provisions that could limit the possibility to implement new pricing schemes such as dynamic pricing, or ToU pricing, concerning the energy commodity component.

In Ireland, the electricity and gas markets are deregulated. Suppliers can set their own tariff regimes, but in the absence of smart meters, these are flat rate tariffs. After the roll-out of smart meters, all suppliers will be obliged by the regulator to have at least one ToU tariff to offer customers post smart meter install and must, on a regular basis, alert customers to the fact that there are ToU tariffs available. There will also be an option for suppliers to introduce dynamically priced tariffs for customers.

5.4.2.4 The Netherlands

The distribution sector in the Netherlands is characterized by eight DSOs, who are subject to incentive-based regulation, that is, price cap. The regulator is responsible for deciding on the final allowed revenues and the tariff structure. The DSO calculates and proposes tariff levels to the regulator for approval.

The tariff structure is based on capacity charges and fixed charges (European Commission, 2015b). The capacity charge, for residential consumers, depends on the maximum contracted connection capacity. These charges include the transport cost of energy, as well as the measuring costs and the connection cost. A unique feature of the Dutch system is that there are no active or reactive energy-based network charges. These capacity-based tariffs are uniform for all residential consumers (meaning that the price in euro/kW is the same). Consequently, no ToU-based distribution tariffs exist in the Netherlands. It is however the case that households carry the largest portion of network costs. This is disproportionate when considering their electricity consumption compared to larger consumer types (source). Finally, there is no network tariff based on injection of electric energy or power.

In the Netherlands, suppliers are allowed to offer dynamic pricing schemes. However, the "Autoriteit Consument en Markt" monitors the energy suppliers and energy prices that are allowed to set maximum prices if energy pricing is set too high, following the monopolistic character of the market. No information is found on regulatory provisions that could limit the possibility to implement new pricing schemes such as dynamic pricing, or ToU pricing, concerning the energy component.

5.4.3 Demand–Response Services

Incentive-based programs trigger demand modification in the occasion of critical events based on contractual arrangements in return for an incentive payment (Dupont, 2015). Although participation is voluntary, falling short on a specific DR usually results in a penalty. In contrast, price-based programs allow the end user to enroll in a dynamic pricing scheme. Voluntary load modifications are based on the user's own economic and rational preferences. In such programs, no penalties are incurred, although the user can be

TABLE 5.5

Incentive-Based Demand-Response Products Accessible for Communities and SMEs

Type	Flanders	Ireland	The Netherlands	United Kingdom
Energy-market services	Yes	Yes	Yes	Yes
Ancillary services	R3-DP	DSU, STAR, Powersave	NV	STOR and FR
Capacity market services	SDR	—	—	DSR CMU

imposed to high electricity prices. Incentive-based demand programs can be divided into different categories (Table 5.5):

Direct load control: Third party is in control of some appliances at the end user's premises (e.g., air conditioners, water boiler, and heat pumps). In the event of system stress, the third party can control those appliances directly in compensation for a previously known participation fee.

Curtailable load: End users are in control of their own appliances. By enrolling into the program, the end user makes the commitment to modify load when a request is received. The gain for the participants can take different forms as bill credits and participation fees. A penalty is given in case the user does not respond to the load signal.

Demand bidding: In demand bidding programs, end users make the commitment to modify load by bidding in the wholesale electricity market. If the bid is cleared, the end user is obliged to reduce his load by the according amount.

These programs can be used by market parties to react to prices in the different electricity markets, that is, (1) energy market in which they can react upon price volatility in day-ahead, intraday, and real-time market; (2) ancillary service markets, such as operating reserves contracted by the system operator, where demand modification can cover unexpected system shortages or excesses; or (3) capacity markets, where demand is used to provide firm capacity in order to cover the expected peak demand. Although there are probably less regulatory provisions for incentive-based DR, the regulatory framework remains relevant for what concerns the availability of DR products, ancillary service and capacity market product requirements which can allow or exclude alternative technologies, are embedded in a regulatory framework. Furthermore, the procurement of these services would benefit from a regulatory framework where actors, roles, and interactions are clearly identified.

5.4.3.1 Energy-Market Products

In general, the regulatory framework of the liberalized market allows consumers to engage in incentive-based programs with respect to the energy market. Although this is rarely observed on a residential level, mainly explained by economic barriers of installing the required metering equipment and ICT infrastructure, this might be different for SMEs with significant electricity consumption. SMEs with flexible demand due to cooling, or other flexible processes, may sell this flexibility to a third party (supplier, producer, and aggregator), which is active on the electricity market.

5.4.3.2 Ancillary Service and CRM Market Products

It is generally accepted that DR can play an important role in the provision of operating reserves of capacity remuneration mechanisms (CRMs). The sizing and allocation of this reserve capacity to specific products are conducted by transmission network operators and approved by regulators. Similarly, CRMs are bound by regulatory provisions. However, results of the survey show that products that allow the participation of aggregated resources are generally available for operating reserve capacity and adequacy services.

In Flanders, specific products are designed by the transmission system operator (TSO) to allow flexibility providers such as aggregated DR to participate in operating reserve (R3-DP) and CRM (strategic reserve) since 2015 strategic demand reserve (SDR). In the Netherlands, no specific products are created for the participation of new flexibility providers, but available operating reserve capacity services such as "emergency power reserves (Noodvermogen, NV)" do facilitate the participation of aggregated DR to provide quick reserve capacity (TenneT, 2013).

In Great Britain, the system operator allows the participation of aggregated DR by means of aggregation in reserve services short term operating reserve (STOR) and frequency response (FR). A specific product STOR Runway is developed to facilitate DR integration (National Grid, 2015). Furthermore, DR is allowed to participate in the future capacity market by means of a demand side response – capacity market unit (DSR CMU) reducing the electricity demand below a predetermined baseline. These units must have a minimum capacity of 2 MW (Ofgem, 2015).

In Ireland, DSM has been and is expected to remain a key operational service to maintain system security for the island. EirGrid currently operates a number of peak demand reduction programs to reduce demand, particularly during the peak periods and tight capacity margins (EirGrid, 2015). A demand-side unit (DSU) consists of one or more individual demand sites that can be dispatched by the TSO as if it was a generator. In addition, the Powersave scheme is implemented to encourage large- and medium-sized customers to reduce their electricity demand on days when total system demand is close to available supply. In return, participating customers are made payments on the basis of kWh reductions achieved during a Powersave Event. Finally, short-term active response (STAR) is a scheme operated by the TSO whereby electricity consumers are contracted to make their load available for short-term interruptions.

5.4.4 Regulatory Practices and Barriers

The deployment of SGs opens up a large spectrum of new services related to flexibility and efficiency. However, in most European countries, small and residential consumers are limited in their ability to offer these services on designated markets, due to regulatory obstructions. These barriers are usually in place with the aim of protecting consumers against unfavorable market conditions or malpractices by competitive players. In the following paragraphs, an overview is given of the main regulatory barriers to active market participation by residential communities and SMEs. Note that this is not an exhaustive list. Some other aspects (e.g., standardization and privacy issues) are also important to consider, but they do not fall within the scope of this chapter.

5.4.4.1 Market Entry

In Europe, one of the most prominent regulatory barriers stems from a lack of market products facilitating participation of demand-side resources, especially in the LV distribution grid. In wholesale electricity markets, fixed charges (e.g., participation fees) and high minimum requirements (e.g., minimum trading volume) make it tough for small players to participate. These requirements are usually embedded in the regulatory framework. The only way to circumvent them is through aggregation in a portfolio, by a third party (e.g., a supplier, retailer, or aggregator).

In the ancillary service market, and in CRMs, this barrier is often even stronger. Operating reserve products, contracted by system operators to maintain the system security, are usually regulated, with very specific technical requirements. These products were initially designed for conventional power plants, and often explicitly exclude DR, or are characterized by technical requirements that exclude DR indirectly. Today, a trend is observed in adapting the product requirements, or creating separate products, to allow the participation of DR, and new flexibility providers.

5.4.4.2 Market Roles

Although most government institutions in Europe take a clear and positive stance on the development of SGs and active demand, there is still a lack of a framework regarding the roles and responsibilities necessary for these changes to happen. Especially when looking at demand participation on the distribution level, there is a lot of uncertainty on what is, and what isn't, allowed to existing and new players.

One example of a challenge to set up such a framework is how to deal with conflicts of interest between parties affected by DR, such as suppliers, balancing responsible parties, aggregators, and network operators. The power system is already strongly intertwined, but DR actions on the distribution level increase this complexity significantly. Examples of questions that need answering include how (and if) suppliers or balance responsible parties (BRPs) should be compensated for potential loss of sales or unexpected imbalances created by demand modifications by a third party.

5.4.4.3 Energy Prices

In most European countries, prices of electricity are not directly regulated or determined by the government or regulator. In a liberalized market, competitive pressure is supposed to keep prices at acceptable levels. ToU pricing, although not frequently applied in practice, is usually allowed as long as it can be accurately and objectively measured.

However, at the same time, most countries have rules in place that protect consumers, by law, against frequent price changes and especially price increases. In several countries, maximum prices for electricity are in place, or the government has the authority to set maximum prices whenever there is a supposed need to. In addition, in the case of variable contract types (meaning that prices are to some extent dynamic and, e.g., subject to indexation based on wholesale market prices), some regions prevent suppliers from frequently changing their prices. In addition, there are transparency requirements (e.g., publishing of the methodology to calculate prices and notification of the regulator). These rules, while established with good intent, sometimes lead to overprotection of consumers and the inability for them to freely decide to purchase DR products based on dynamic pricing, simply because these products are not allowed to exist.

5.4.4.4 Network Tariffs and DSO Remuneration

The existing regulation of electricity distribution tariffs in most member states is still consistent with the traditional features of the distribution business. As a result, most tariff methodologies in the EU are based on active energy off-take and fixed charges. Although this way of charging for network costs is easy for all parties involved, it can be argued that it is not very efficient. Network costs are more related to the capacity of installed cables and transformers than to energy transported. Therefore, the tariffs do not give an incentive to users to behave in a way that is most cost efficient for the grid as a whole. In most countries, there are discussions and plans to reform the grid tariffs and to give more weight to capacity-based tariff drivers. This is consistent with the idea of development of demand-side flexibility toward commercial market participants, which is also strongly motivated by the cost-efficiency argument. So far, the Netherlands is the only country in Europe to completely abolish energy-based tariffs.

However, in their quest for efficient grid tariffs, many regulators are implementing charges that are based on capacity, but are not truly efficient. For example, the grid tariffs in the Netherlands are based on the contracted capacity of the user (i.e., the capacity of their connection). Although related to the initial cost of connecting the user, this tariff driver does not account for the actual usage of the grid. Users who have the same connection capacity but widely different consumption profiles (e.g., peak consumption vs. off-peak consumption) still pay the same rate. A second example is the new prosumer tariff in Flanders. This tariff is based on the capacity of the PV system inverter, and therefore it discourages investment in PV altogether, even when it is actually beneficial to the grid. It does not provide incentives for active management of demand and assumes that all systems with the same size have the same impact on network costs. These examples illustrate how old and new forms of tariff design still limit the incentive for users to actively manage their consumption and generation. In some cases, the grid tariffs even incentivize quite the opposite. As a result, investment in SG solutions and active demand may slow down.

5.5 Conclusions

The results of this study show that the emission reductions by the deployment of SG technologies (e.g., ICT hardware) are relatively large. Moreover, a larger penetration of renewable energy, such as PV and heat pumps, can lead to additional CO_2 emissions. However, the choice of hardware components should be done thoughtfully, so that the additional benefits at least offset the added CO_2 emissions. However, it should be noted that the emission savings from smart control are relatively smaller than renewable energy production and come at the expense of increases in electricity consumption and grid interaction.

In all countries considered in this study, retail electricity prices are found to be primarily flat (i.e., fixed over time per unit consumed). Although most countries have some forms of ToU products in place, these are usually limited to day and night schemes or seasonal rates, which are communicated to the user far in advance. In some countries, dynamic pricing is found to be severely limited, even in cases where consumers voluntarily wish to participate. In Belgium, for instance, prices in variable contracts can only be indexed up to four times per year, at the start of each trimester. In other countries, such as the Netherlands or the United Kingdom, the regulatory framework gives more leeway. However, implementation remains limited due to techno-economic barriers, such as affordable metering and control.

While flat prices do incentivize low consumption (and by extension fewer emissions), they do not incentivize the shifting of consumption toward times when generation costs and/or emissions are lowest. As a result, these pricing schemes do not incentivize the use of SG technologies. It would therefore be advisable for policy makers to develop a regulatory framework that, at the very minimum, allows the development of DR products based on dynamic pricing, providing sufficient transparency and protection mechanisms against excessive price changes.

With respect to distribution grid tariffs, it is found that these are usually flat (although day and night schemes do exist) and based on energy off-take, with an added fixed component related to metering and administrative costs. The Netherlands is an exception in this regard, being the only country in the NWE to charge a tariff completely based on the contracted capacity of the user. In general, flat and energy-based tariffs are not reflective of the underlying costs of grid operators, which are more related to capacity. In addition, they often incentivize a fit-and-forget approach, making investment in SG infrastructure less attractive.

Despite the fact that some countries have made reform efforts toward introducing capacity-based tariffs, the proposed solutions do not remove these shortcomings. Tariffs based on the capacity of the connection (e.g., in the Netherlands) are not reflective of costs due to actual grid usage. Tariffs based on the capacity of the PV system (e.g., in Belgium) are not reflective of the actual costs caused by local generation and discourage PV investment even when it is beneficial to the grid. In neither of the countries considered are tariffs based on actual peak usage, or usage during system peaks. Yet, these types of tariffs may encourage more efficient usage of the grid, which would be in favor of SG technologies. The roll-out of smart meters plays an important role in this regard, as it removes a long-standing technical barrier to improvement of the grid tariffs.

In general, the market design does not yet include an adequate framework on the procurement of DR services from users connected to the distribution grid. However, it can be observed that all regions allow participation of DR in network services, or in CRMs. Some countries developed specific products that favor DR (e.g., R3-DP as an operating reserve product in Belgium). Other countries adapted the product requirements in a way that participation of demand is allowed (e.g., Noodvermogen in the Netherlands). Aggregators play a key role in encouraging the participation of distributed demand-side flexibility providers, pooling small pieces of flexibility into marketable products. Note that only SMEs with fairly high electricity consumption are now participating in these products and that much work is yet to be done in order to bring this trend to the residential level.

At the same time, it can be observed that existing actors in the electricity market are often not strictly prohibited from developing DR products (e.g., when they are based on incentive payments). However, the playing field is characterized by uncertainty, due to a lack of clear rules and responsibilities. DR services can have an impact on several market participants, and many questions on the interactions between the system operators, aggregators, and suppliers remain unanswered. The introduction of flexibility on the level of communities and SMEs requires a clear and complete definition of the roles of existing and new market actors in the development and implementation of new DR products. More specifically, the current set of tasks attributed to the TSO, the DSOs, the supplier, the generators, the BRPs, and the balancing service providers (BSPs) should be reviewed. Clear and fair contractual and communication requirements need to be developed, both for existing and new parties. This is especially needed in the relationship between the BRP (often a generator and supplier) and the aggregator or BSP. In addition to bilateral flexibility services, centralized trading platforms can be developed for the provision and procurement of flexibility in the form of standardized products.

Because SG technologies and their control algorithms heavily depend on data (e.g., data related to day-to-day consumption), smart meters, or at least the data provided by them, play a key role in facilitating their deployment. The advantages of smart meters include frequent and accurate data collection of bidirectional electricity flows, as well as remote communication with other appliances. In order to fully exploit their potential, well-defined measurement and verification protocols are needed, as well as standards for interoperability.

All countries considered in this survey have either decided to roll out smart meters or are still in the process of developing a roll-out strategy. In Flanders, for instance, the minister of energy confirmed her intentions to proceed with a roll-out from 2019 onward, but a definite political decision will be made by the end of 2015. The other regions have already made specific plans for roll-out within the next 5 years. There is, however, a difference in the approach: whereas most regions choose a regulated market model, with the DSO as the owner and operator of the meter and its data, the United Kingdom has chosen for a commercial model, in which the supplier is responsible for roll-out, and data collection happens through a central hub.

Notwithstanding the benefits smart meters can provide, it should be noted that the ownership model of these meters, and especially of their data, may strongly affect the extent to which new demand-side services can be designed. One downside resulting from having a regulated metering model, in which ownership lies with the DSO and metering functionalities are uniform and based on predefined standards, is that this is a one-size-fits-all solution. For many consumers, if not most, the data provided by smart meters may be partly unnecessary. For others, the data may be insufficient. Having a competitive metering model allows for tailored solutions. However, because meters and their data have high strategic value, a competitive ownership model can negatively impact competition in the market. Which model to adopt, however, remains a political choice. In the end, the consumer should play a central role in the debate. Some barriers to the development of SG solutions are less obvious but may nonetheless have an effect that should not be underestimated. An example of this is the current way in which most EU countries regulate the aggregate costs and revenues of the DSOs. Traditionally, regulators used a cost-plus approach, meaning that all costs were subject to regulatory approval, after which a reasonable return on investment was granted. Now, most member states have switched toward incentive-based approaches. This usually comes down to implementation of price or revenue caps, which are based on past revenues, inflation, evolution of input price indices, and an efficiency term. Operators are allowed to keep the difference between the revenue cap and their actual costs, and as such, they receive an incentive toward efficient operation and investment. At least that is the main objective of this regulatory mechanism. The downside is that operators simultaneously receive an incentive to cut down on expenses that lead to long-term benefits (e.g., R&D costs). Instead, they may prefer to substitute these expenses with solutions that are less costly in the short term but may also be less beneficial in the long term. In other words, this type of regulation, without additional measures, may slow down SG development.

Another aspect that is often overlooked, but is nonetheless important from the side of the consumer, is privacy. The implementation of SGs, and especially smart electricity meters, implies a large increase in the amount of data gathered and its exchange between different, sometimes commercial, players. Metering data can give very detailed insight in the consumption patterns of consumers, making it, for example, possible to determine whether or not they are at home. It goes without saying that data security is crucial and that there is sufficient transparency toward consumers, such that they are aware of who has access to their personal data. The European legal framework on privacy is quite clear in pointing out that data processing of personal data is not allowed without personal consent, except for a limited number of occasions. Given the fact that processing of consumption data on the level of an individual user (instead of, for instance, street level) has never been a requirement for secure network operation, it can be argued that the data provided by a smart meter has primarily commercial merits. Therefore, without a clear legal framework on privacy and data protection specifically in the context of SGs, consumers may feel reluctant to participate in new products and services.

Acknowledgments

This study has been implemented within the GREAT (Growing Renewable Energy Applications and Technologies) project funded by the European

INTERREG IVB North-Western Europe Programme. Nele Witters was financed by FWO (Research Foundation Flanders).

References

Adamec, M., P. Pavlatka, and O. Stary. 2011. Costs and benefits of smart grids and accumulation in Czech distribution system. *Energy Procedia* 12: 67–75.

Amin, M. and J. Stringer. 2008. The electric power grid: Today and tomorrow. *MRS Bulletin* 33(4): 399–407.

Basso, G., N. Gaud, F. Gechter, V. Hilaire, and F. Lauri. July 2013. A framework for qualifying and evaluating smart grids approaches: Focus on multi-agent technologies. *Smart Grid and Renewable Energy* 2013: 333–347.

Battaglini, A., J. Lilliestam, A. Haas, and A. Patt. 2009. Development of SuperSmart grids for a more efficient utilisation of electricity from renewable sources. *Journal of Cleaner Production* 17(10): 911–918.

Blumsack, S. and A. Fernandez. 2012. Ready or not, here comes the smart grid! *Energy* 37(1): 61–68.

Connor, P.M., P.E. Baker, D. Xenias, N. Balta-Ozkan, C.J. Axon, and L. Cipcigan. 2014. Policy and regulation for smart grids in the United Kingdom. *Renewable and Sustainable Energy Reviews* 40: 269–286.

Cunjiang, Y., Z. Huaxun, and Z. Lei. 2012. Architecture design for smart grid. *Energy Procedia* 17: 1524–1528.

Dupont, B. 2015. *Residential Demand Response Based on Dynamic Electricity Pricing: Theory and Practice.* KU Leuven—Faculty of Engineering Science, Heverlee, Belgium.

EPRI. 2008. The Green Grid. *Manager* 3(3): 1–64.

Eurelectric. 2012. *The Smartness Barometer—How to Quantify Smart Grid Projects and Interpret Results.* A EURELECTRIC paper Union of the Electricity Industry—EURELECTRIC aisbl. Brussels, Belgium.

European Commission. 2010. *Energy 2020. A Strategy for Competitive, Sustainable and Secure Energy.* COM(2010) 639 final. European Commission. Brussels, Belgium doi:COM(2010) 639.

European Commission. 2014. *Cost-Benefit Analyses & State of Play of Smart Metering Deployment in the EU-27.* SWD(2014) 188 final, accompanying the document *COM(2014) 356 final.* European Commission, Brussels, Belgium.

European Commission. 2015a. *Energy Union Package—A Framework Strategy for a Resilient Energy Union with a Forward-Looking Climate Change Policy.* COM(2015) 80 final. European Commission, Brussels, Belgium.

European Commission. 2015b. *Study on Tariff Design for Distribution Systems.* Directorate-General for Energy Directorate B—Internal Energy Market. European Commission, Brussels, Belgium.

Faruqui, A., D. Harris, and R. Hledik. 2010. Unlocking the €53 billion savings from smart meters in the EU: How increasing the adoption of dynamic tariffs could make or break the EU's smart grid investment. *Energy Policy* 38(10): 6222–6231.

Faruqui, A., R. Hledik, and S. Sergici. 2009. Piloting the smart grid. *Electricity Journal* 22(7): 55–69.

Faruqui, A., D. Mitarotonda, L. Wood, A. Cooper, and J. Schwartz 2011. The costs and benefits of smart meters for residential customers. White Paper, July.

Farzaneh, H., A. Suwa, C.N.H. Dolla, and J.A.P. De Oliveira. 2014. Developing a tool to analyze climate co-benefits of the urban energy system. *Procedia Environmental Sciences* 20: 97–105.

Fouquet, D. and T.B. Johansson. 2008. European renewable energy policy at crossroads—Focus on electricity support mechanisms. *Energy Policy* 36(11): 4079–4092.

Frischknecht, R., N. Jungbluth, H. Althaus, G. Doka, R. Dones, T. Heck, S. Hellweg et al. 2007. Overview and Methodology. Data v2.0 (2007). Ecoinvent report No. 1. Dübendorf, Switzerland.

Gellings, C., 2011. Estimating the costs and benefits of the smart grid: A preliminary estimate of the investment requirements and the resultant benefits of a fully functioning smart grid. Electric Power Research Institute (EPRI), Technical Report (1022519).

Görbe, P., A. Magyar, and K.M. Hangos. 2012. Reduction of power losses with smart grids fueled with renewable sources and applying EV batteries. *Journal of Cleaner Production* 34: 125–137.

Hidayatullah, N.A. 2011. Analysis of distributed generation systems, smart grid technologies and future motivators influencing change in the electricity sector. *Smart Grid and Renewable Energy* 2(3): 216–229.

Hischier, R., B. Weidema, H. Althaus, C. Bauer, G. Doka, R. Dones, R. Frischknecht et al. 2010. Implementation of Life Cycle Impact Assessment Methods Data v2.2 (2010). Ecoinvent Report No. 3. Dübendorf, Switzerland.

Hledik, R. 2009. How green is the smart grid? *Electricity Journal* 22(3): 29–41.

International Energy Agency 2013. Technology Roadmap: Smart Grid. doi: 10.1007/SpringerReference_7300.

Langheim, R., M. Skubel, X. Chen, W. Maxwell, T.R. Peterson, E. Wilson, and J.C. Stephens. 2014. Smart grid coverage in U.S. newspapers: Characterizing public conversations. *Electricity Journal* 27(5): 77–87.

Lewis, P. 2013. *Smart Grid Global Impact Report 2013*. VaasaETT, San Francisco, CA. http://www.smartgridimpact.com/impact_report_2013.pdf. Accessed on November 26, 2015.

Livieratos, S., V.E. Vogiatzaki, and P.G. Cottis. 2013. A generic framework for the evaluation of the benefits expected from the smart grid. *Energies* 6(2): 988–1008.

Luthra, S., S. Kumar, R. Kharb, Md.F. Ansari, and S.L. Shimmi. 2014. Adoption of smart grid technologies: An analysis of interactions among barriers. *Renewable and Sustainable Energy Reviews* 33: 554–565.

Moretti, M. et al. 2017. A systematic review of environmental and economic impacts of smart grids. *Renewable and Sustainable Energy Reviews*, 68, 888–898. Available at: http://dx.doi.org/10.1016/j.rser.2016.03.039.

National Grid. 2015. *STOR Runway*. Guidance document December 2015. Gallows Hill, Warwick, U.K.

Ofgem. 2015. The Capacity Market (Amendment) Rules 2015.

Pérez-Arriaga, I.J., S. Ruester, S. Schwenen, C. Battle, and J.-M. Glachant. 2013. From distribution networks to smart distribution systems: Rethinking the regulation of European electricity DSOs.

Personal, E., J.I. Guerrero, A. Garcia, M. Peña, and C. Leon. 2014. Key performance indicators: A useful tool to assess smart grid goals. *Energy* 76: 976–988.

Peterson, S.B., J.F. Whitacre, and J. Apt. 2010. The economics of using plug-in hybrid electric vehicle battery packs for grid storage. *Journal of Power Sources* 195(8): 2377–2384.

Pratt, R.G., P.J. Balducci, C. Gerkensmeyer, S. Katipamula, M.C.W. Kintner-Meyer, T.F. Sanquist, K.P. Schneider, and T.J. Secrest. January 2010. The smart grid: An estimation of the energy and CO_2 benefits, pp. 1–172.

Ruester, S., S. Schwenen, C. Batlle, I. Pérez-Arriaga. 2014. From distribution networks to smart distribution systems: Rethinking the regulation of European electricity DSOs. *Utilities Policy*, 31, 229–237.

Simoes, M.G., R. Roche, E. Kyriakides, S. Suryanarayanan, B. Blunier, K.D. McBee, P.H. Nguyen, P.F. Ribeiro, and A. Miraoui. 2012. A comparison of smart grid technologies and progresses in Europe and the U.S. *IEEE Transactions on Industry Applications* 48(4): 1154–1162.

Sullivan, M. and J. Schellenberg. April 2011. Smart grid economics: The cost-benefit analysis. *Renew Grid*, pp. 1–4.

Sun, Q., X. Ge, L. Liu, X. Xu, Y. Zhang, R. Niu, and Y. Zeng. 2011. Review of smart grid comprehensive assessment systems. *Energy Procedia* 12: 219–229.

Tekiner-Mogulkoc, H., D.W. Coit, and F.A. Felder. 2012. Electric power system generation expansion plans considering the impact of smart grid technologies. *International Journal of Electrical Power & Energy Systems* 42(1): 229–239.

Van Dam, S.S., C.A. Bakker, and J.C. Buiter. 2013. Do home energy management systems make sense? Assessing their overall lifecycle impact. *Energy Policy* 63: 398–407.

Wang, W., Y. Xu, and M. Khanna. 2011. A survey on the communication architectures in smart grid. *Computer Networks* 55(15): 3604–3629.

Yadoo, A. and H. Cruickshank. 2012. The role for low carbon electrification technologies in poverty reduction and climate change strategies: A focus on renewable energy mini-grids with case studies in Nepal, Peru and Kenya. *Energy Policy* 42: 591–602.

Yuan, J. and Z. Hu. 2011. Low carbon electricity development in China—An IRSP perspective based on super smart grid. *Renewable and Sustainable Energy Reviews* 15(6): 2707–2713.

Zakariazadeh, A., S. Jadid, and P. Siano. 2014. Economic-environmental energy and reserve scheduling of smart distribution systems: A multiobjective mathematical programming approach. *Energy Conversion and Management* 78: 151–164.

6

Renewables Optimization in Energy-Only Markets

Patrícia Pereira da Silva and Nuno Carvalho Figueiredo

CONTENTS

6.1 Introduction

The promotion of renewable energy source for electricity (henceforth referred to as RES-E) by the European Union (EU) aims to reduce dependency on imported fossil fuels and greenhouse gas (GHG) emissions, resulting in the successful deployment of RES-E generation in Europe (European Union 2009a). This has been achieved through a set of energy policies, comprising, among others, strong financial instruments, like feed-in tariffs, feed-in premia, fiscal incentives, and tax exemptions (Meyer 2003; Jager et al. 2011).

The changes in the European electricity systems are profound and ongoing. New challenges arise from the high-level penetration of RES-E, both in the technical sense and in the market design, due to the known RES-E intermittency and nondispatchability (Benatia et al. 2013).

Simultaneously, electricity markets in Europe are being restructured in face of a number of European policies intending to guarantee the supply of electricity, reduce costs, foster competition, ensure security of supply, and protect the environment (European Union 2009b). Alongside, unbundling and privatization of the electricity supply industry has been achieved in most of the EU Member States, together with the creation of independent national regulatory agencies, introducing competition at the different market levels (da Silva 2007). Energy-only markets remunerate electrical energy based on the traded volume and price. Therefore, increasing RES-E creates a depression in spot electricity prices due to the merit-order effect of zero marginal cost bidding and diminishes the available load for the remaining nonzero bidding technologies (Traber and Kemfert 2011). The size of this residual load (Henriot and Glachant 2013) sets the electricity spot market price and provides the main income to electricity suppliers. Thus, one of the fundamental issues affecting electricity markets is the integration of RES-E and the associated impact on price signals for investment in the electricity system. In parallel, the European Union Emissions Trading Scheme (EU ETS), based on the "cap-and-trade" principle, emerged to be the cornerstone of the European Union's policy to combat climate change and its key tool for reducing industrial GHG cost effectively (Freitas and da Silva 2015). Among the several industries covered by the scheme, the electricity sector is the largest one. Launched in 2005, the implementation of the EU ETS was set to run in three phases: the first (pilot phase) ranging from 2005 to 2007, the second from 2008 to 2012, and now in its third phase running from 2013 to 2020. Nevertheless, the collapse of the CO_2 price weakens the link between the carbon market and the electricity market, consequently putting at risk the policy goals associated with carbon pricing (da Silva et al. 2015, 2016; Moreno and da Silva 2016) and, thus, leaving increased relevance for the role of RES-E.

In this chapter, an analysis of the main concerns in integrating RES-E into the spot electricity markets is provided. The influence of high-level RES-E in "energy-only" electricity markets is discussed, highlighting its optimization through market integration. In Section 6.2, an overview of the experienced growth in RES-E generation is delivered, followed in Section 6.3 by the description of two main concerns of high RES-E penetration currently in the minds of many stakeholders in the electricity sector. In Section 6.4, the analysis of the issues, challenges, and strategies that the European electricity sector faces with the integration of high levels of RES-E is presented and discussed. The RES-E optimization through regional market integration is then highlighted in Section 6.5, as it is considered one of the most important items in RES-E market integration and consequent optimization. Final remarks are presented in Section 6.6.

6.2 Growth of RES-E

The world demand for energy calls for increasing sustainable energy systems. "Sustainable" means, in this context and according to Brundtland's report (Brundtland 1987), energy that does not jeopardize future generations, a reality that can be accomplished through renewable energy sources.

In line with this, the development of renewable energy technologies aims to improve energy security, decrease the dependency on fossil fuels, and reduce greenhouse gas emissions.

Europe's ambitious target of 20% renewable energy sources in 2020 (or 33% renewable energy sources for electricity) prompted several member states to propose highly attractive support mechanisms. Denmark, Germany, Portugal, Spain, Italy, Ireland, and Belgium, for example, have seen their share of renewable energy sources, mainly in wind and solar, increase drastically in a few years.

Among all renewable energy sources, wind and solar were the ones subject to the strongest research and development, based on clusters established in some regions of Europe. All these efforts required financial instruments like feed-in tariffs, feed-in premia, fiscal incentives, tax exemptions, and others (Meyer 2003; Klessmann et al. 2008; Jager et al. 2011; Amorim et al. 2013). These financial instruments provided an initial incentive to invest in nonmature RES-E technologies. However, with time, wind and solar power became mature and investment costs decreased to levels where these instruments are obsolete. In fact, the financial burden of RES-E incentive policies is significant and are being reviewed in Europe. Germany and Spain, for instance, took actions reducing RES-E financial support (Moreno and Martínez-Val 2011; Diekmann et al. 2012).

One of the most successful examples of RES-E incentive policies can be found in Denmark, where a partnership between public and private institutions was established (Danish Energy Authority 2007). By 1972, Denmark did not have significant wind power, which after a strong energy policy shift managed to reach 20% RES-E share in 2008 (Lund 2010; Lund et al. 2013). Since then, the RES-E share in Denmark continued to rise, reaching, in 2015, 41.4% of wind power and 13.8% of essentially biomass (Figure 6.1). This level of RES-E is possible due to the cross-border interconnections that allow electricity trading in the Nord Pool and smooths production profiles with the use of neighboring pumped storage hydro plants. The Danish 50% target for wind power production can only be achieved with strong interconnected electricity markets (Benatia et al. 2013).

Other European countries also pursued the same route of RES-E deployment. Both Iberian countries had an outstanding increase in wind power, while only in Spain there was significant development in solar power. Furthermore, the hydropower generation share historically high in Iberia as seen in Figures 6.2 and 6.3. *Energiewende* in Germany is the policy shift that prescribed the nuclear phase-out and the replacement of fossil generation with RES-E. Figure 6.4 will illustrate that this policy has been quite successful in deploying wind, solar, and biomass; Germany has currently the largest wind and solar power in Europe with 40.5 and 38.2 GW of installed capacity, respectively (BP 2015). Similar RES-E developments are scheduled throughout Europe, depending on country-specific energy policies and financial incentives available. For example, as shown in Figure 6.5, in the

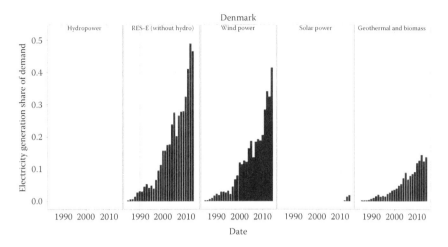

FIGURE 6.1
Hydropower, RES-E, wind, solar, and geothermal/biomass electricity generation shares evolution in Denmark. (From BP, Statistical Review of World Energy 2015, 2015, http://www.bp.com/en/global/corporate/energy-economics/statistical-review-of-world-energy.html. Accessed March 31, 2016.)

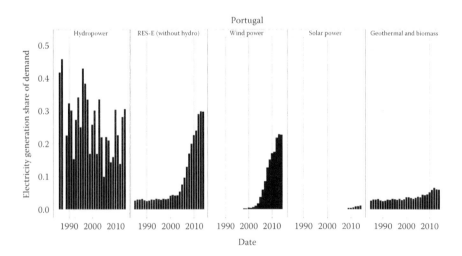

FIGURE 6.2
Hydropower, RES-E, wind, solar, and geothermal/biomass electricity generation shares evolution in Portugal. (From BP, Statistical Review of World Energy 2015, 2015, http://www.bp.com/en/global/corporate/energy-economics/statistical-review-of-world-energy.html. Accessed March 31, 2016.)

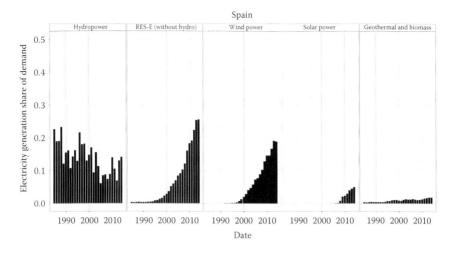

FIGURE 6.3

Hydropower, RES-E, wind, solar, and geothermal/biomass electricity generation shares evolution in Spain. (From BP, Statistical Review of World Energy 2015, 2015, http://www.bp.com/en/global/corporate/energy-economics/statistical-review-of-world-energy.html. Accessed March 31, 2016.)

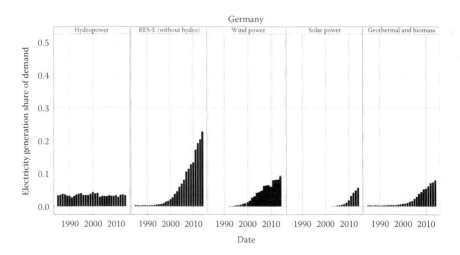

FIGURE 6.4

Hydropower, RES-E, wind, solar, and geothermal/biomass electricity generation shares evolution in Germany. (From BP, Statistical Review of World Energy 2015, 2015, http://www.bp.com/en/global/corporate/energy-economics/statistical-review-of-world-energy.html. Accessed March 31, 2016.)

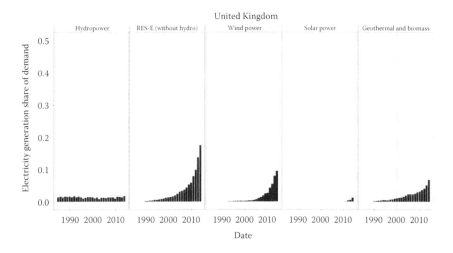

FIGURE 6.5
Hydropower, RES-E, wind, solar, and geothermal/biomass electricity generation shares evolution in the United Kingdom. (From BP, Statistical Review of World Energy 2015, 2015, http://www.bp.com/en/global/corporate/energy-economics/statistical-review-of-world-energy.html. Accessed March 31, 2016.)

UK wind power generation share grew to 9.4%, however without significant solar power development.

The EU 2030 targets a RES-E share increase to 45%, revealing that RES-E still needs to grow, displacing technologies with higher greenhouse gas emissions and thus contributing to its desired reduction. Impacts of this high level of RES-E penetration on electricity markets are discussed in the following sections, starting with the effects on the existing energy-only markets and related influence on utility business, followed by some strategies to facilitate the transition to a more sustainable electricity system.

6.3 "Merit-Order Effect" and the "Missing Money Problem"

Electricity trading in Europe is currently based on several types of markets: exchanges or spot markets, bilateral and over-the-counter markets, ancillary services markets, and retail markets. Presently, electricity exchanges in Europe trade volumes of electricity at a clearing price, matching supply and demand. All market agents bidding lower than the clearing price trade their bidding volumes at that price. These exchanges have day-ahead sessions for each of the day period (usually for each of the 24 hours) and intraday sessions to provide a first level for the electrical system balance.

The electricity market price clearance is done for a specific geographical area, which depends not only on national borders, but also in some cases on internal transmission capacity, reflecting electricity flow constraints and allowing for distinct price signals in each area (e.g., Sweden with four bidding areas). In Europe, spot electricity market bidding areas are then joined through a market coupling/splitting mechanism where bidding areas with lower prices export electricity to markets with higher prices through the interconnections. If the interconnection capacity is large enough to accommodate the exported electricity flows (without congestion), then the price is the same in both markets, otherwise market splitting occurs and two regional market prices are cleared (EPEX et al. 2010).

On the supply side, the so-called merit-order of generators depends on the marginal costs of each market agent bidding in the spot electricity market. These marginal costs of market agents depend mainly on the generation technology in their electricity production portfolio and related operational costs (Eydeland and Wolyniec 2003). Each generating plant operational cost presents several components like fuel, variable consumables, variable maintenance, emissions, and transmission costs. Generally, in the bottom of the supply curve one can find market agents bidding electricity produced with low marginal cost technologies, like nuclear or hydro. This is the also the case of renewable generation technologies with high capital costs and small operational costs, which will produce as much electrical energy as the applicable renewable resource available (Klessmann et al. 2008). Therefore, electricity spot prices are significantly dependent on the available renewable electrical energy in the market, given that renewable power comes first in the merit-order, lowering spot electricity prices and potentially causing zero, or even negative, price periods in the case when demand is fully covered (Felder 2011; Schaber et al. 2012).

Confirmation of this is obtained through the analysis of data extracted from the Iberian electricity spot market (OMIE), from July 1, 2008, to March 15, 2014, where the volume of bids at zero price is found to be positively correlated with the available RES-E power generation, as seen in Figure 6.6. Clearly, the spot electricity price is also correlated with the volume of bids at zero price, however, negatively, with significant amount of market periods with zero spot electricity price (Figure 6.7), confirming the statements of Schaber et al. (2012), Felder (2011), and Edenhofer et al. (2013).

Renewable power bids shift the aggregated supply curve to the right and displace high marginal cost generation out of the merit order. This, as mentioned earlier, is the so-called merit-order effect, causing a reduction in the spot electricity price and reducing the load available for conventional power, or the so-called residual load (Sensfuß et al. 2008; Felder 2011; Henriot and Glachant 2013). The residual load is positively correlated with the spot electricity price, as observed for the OMIE in Figure 6.8. In Figure 6.9, we can detect that the hour with the highest RES-E generated in Iberia in the data sample extracted from the OMIE was on January 28, 2014, hour 20. Considering the

2008-07-01 to 2014-03-15

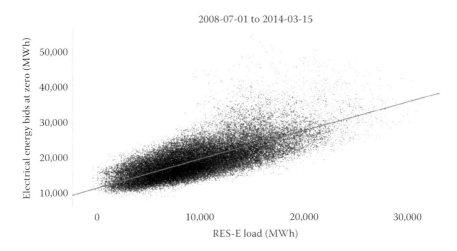

FIGURE 6.6
OMIE electrical energy bids at zero vs. renewable power generation.

2008-07-01 to 2014-03-15

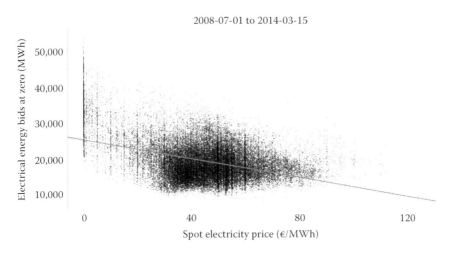

FIGURE 6.7
OMIE electrical energy bids at zero vs. spot electricity price.

aggregated supply curves with, and without, the RES-E bids, it is possible to compute the merit-order effect, which for this hour alone amounted to 2.1 million Euros.

Felder (2011) actually stated that by providing incentives to "out-of-market" technologies, such as most renewables, spot electricity prices would fall to zero. Lower spot electricity prices* are often used to justify the incentives

---

* For example, each GWh of RES-E predicted in German-Austria Würzburg et al. (2013) reported €1/MWh decrease in spot electricity price.

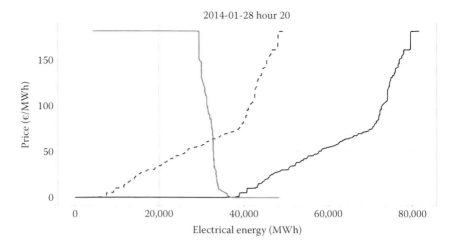

FIGURE 6.8
OMIE residual electrical energy vs. spot electricity price.

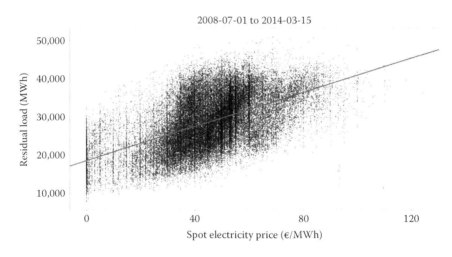

FIGURE 6.9
OMIE aggregated demand and supply curves (with RES E bids—solid and without RES-E bids—dashed).

provided to RES-E; however, they create a number of challenges related with the investment signals and capital cost recovery. Additionally, wealth fails to shift from producers to consumers (Sensfuß et al. 2008; Gelabert et al. 2011; Würzburg et al. 2013), as in most cases savings are not obtained by consumers due to the inclusion of renewable incentive costs in their electricity bills.

Additional concerns and challenges of high generation shares of RES-E are reported both in the technical sense and in the market design. On the technical sense, it is possible to list the following: generation variability and uncertainty, adequate transmission capacity, flexibility and standby of dispatchable generation, electrical system regulation and frequency control, demand-side response, RES-E curtailment, energy storage, adequate transmission grid, and cross-border interconnections (Mauritzen 2010; Nicolosi 2010; Edenhofer et al. 2012; Lynch et al. 2012). Concerning the market design, one can enumerate electricity market integration, the cost allocation of transmission grid and cross-border interconnections, intraday and reserve power markets, RES-E financial support schemes, and capacity support mechanisms (Nicolosi 2010; MIT Energy Initiative 2011; Batlle et al. 2012; Benatia et al. 2013).

Vis-à-vis market design, the reduced residual load and the depressed spot electricity prices, along with the technical challenges and costs of peaking conventional thermal power plants, are currently stressing utilities income. A revenue reduction of 60% for conventional power plants in regions with high RES-E penetration is reported, making capital cost recovery problematic (Schaber et al. 2012; Würzburg et al. 2013). Moreover, in the presence of barriers to exit,* conventional power producers remain available as market agents, further contributing for system electricity surplus, thinning costs to a level where plant reliability may pose an issue (Nelson et al. 2015). Higher volatility can be expected with low plant reliability, which under an "energy-only" electricity market could provide adequate price signals to stakeholders. Nevertheless, high volatility and price caps conflict with these signals, rendering investment in new plant unattractive in the long run. The "missing money problem" of an energy-only electricity market arises when the market fails to provide incentives to sustain adequate generation capacity. Balancing markets and ancillary services, usually run by system operators, if suitably remunerated, might mitigate this issue by providing additional income to generators that are able to deliver these types of services to the grid† (Cramton and Stoft 2006; Edenhofer et al. 2013; Newbery 2015).

The "missing money problem" not only impacts conventional generation but also affects RES-E market integration. If RES-E is exposed to market risks without the known financial support, given the depressed short-term marginal pricing from an "energy-only" electricity market, capital cost recovery would be problematic. Thus, investment in RES-E can also be at risk depending on the future developments of financial incentives and electricity market integration of RES-E. Large amounts of RES-E might only be financially sustainable if incentives are kept and market integration and design is carefully considered. Given the EU targets of RES-E expansion to

---

* Exit barriers, originating from policy or economic reasons, means retiring plants from the electricity market not mothballing (Nelson et al. 2015).
† Balancing services can consist of primary reserve, secondary reserve, automatic generation control, voltage and frequency control, and black start.

45% generation share, further spot electricity price reductions will be seen, aggravating the missing money problem. Edenhofer et al. (2013) summarize three possible causes for the "missing money problem": capped spot prices during scarcity events, low spot electricity prices to sustain existing capacity, and investors discouraged by high price volatility and risks. A generation adequacy problem arises, given the absence of new capacity deployment (Cramton and Stoft 2006).

The challenges faced by the integration of high levels of RES-E require the introduction of additional strategies in the European electricity sector. These are discussed in the following section.

6.4 Market Integration of High Level RES-E

The market integration of RES-E is currently a hot topic and it is being addressed by an increasing number of researchers. The large penetration of RES-E in some of the European electricity markets created a set of challenges, both in a technical and a market perspective. As already unveiled in the previous section, the high-level RES-E deployment caused market failures and distrust in the energy-only electricity markets implemented throughout the EU (Edenhofer et al. 2013). The generation mix is not market driven, creating a nonsustainable financial situation for both utilities and consumers, the former with impaired revenues not being able to recover investments and the latter having to support high value subsidy schemes for RES-E. The associated costs of the financial support mechanisms to RES-E raise some concerns. With RES-E technologies becoming mature, a gradual reduction of subsidies would be expected, due to the reduction investment costs, increasing competitiveness, and the subsidy expiration of older units. Germany and its "Energiewende" is in the forefront, aiming to replace nuclear and coal power generation in one go, nevertheless with a demanding cost containment exercise (Würzburg et al. 2013; Hirschhausen 2014).

Two key expressions were introduced by Henriot and Glachant (2013) considering the integration of RES-E in the electricity markets and associated risks: *melting-pot* and *salad-bowl*. The former exposes RES-E to the same rules as any other conventional generator capable of controlling dispatch (performing as any market agent bidding volumes of electricity at a price for each market period and subject to imbalancing charges by noncompliance to deliver scheduled volumes of electrical energy), while the latter could accommodate two different sets of rules, one for dispatchable units and another for nondispatchable units. It is argued that RES-E particularities are inadequate for spot market bidding, since there is no control on the available renewable resource (therefore, no control on the electricity volumes fed into the system), the prediction of future volumes of electricity generation is limited (high risk of exposure to imbalancing charges), and, with low marginal cost pricing, investment costs might not be recovered (additionally, there are no incentives to invest in new RES-E and conventional power plant capacity). Moreover, Batlle et al. (2012) endorse that the market power of incumbents would increase when owning RES-E and conventional power simultaneously, by adopting a strategic behavior. However, without price signals, RES-E might not have incentives to optimize operational costs and the existing price signals, dependent on the residual load, are not adequate to sustain conventional power (Klessmann et al. 2008).

Flexibility of the electricity system is paramount in obtaining an efficient electricity market incorporating a high level of RES-E. A number of proposals are laid down in the literature to disentangle the RES-E market integration issues, discussed earlier, and introduce the required flexibility, as listed here:

- A premium system allows RES-E compensation above the spot electricity market, limiting market risk and allowing investment cost recovery. RES-E would be subject to the same market rules and risks of the other agents with conventional dispatchable power, including imbalancing charges applicable to deviations from programmed electrical energy. Thus, forecasting RES-E generation is of the upmost importance, improving system predictability and minimizing imbalance costs (Klessmann et al. 2008). A similar system, spot market plus a premium with a cap and a floor, is already implemented in Spain as an option to agents with more than 1 MW, subject to all market rules, with the exception of mandatory secondary reserve market participation and the reactive regime remaining the same (Rivier Abbad 2010). Furthermore, Nicolosi (2010) states that this premium system for RES-E would limit negative pricing as is the case in Germany.

- Improve demand-side response, including households and industry load management, and in a foreseeable future electrical vehicle smart charging. This would make the electrical system more flexible to cope with RES-E intermittency (Benatia et al. 2013).

- Development of storage technology in addition to hydro-pumped-storage, and growing storage installed capacity to allow the use of electricity surplus when there is abundant renewable resource, increasing RES-E utilization (Benatia et al. 2013). The use of compressed air energy storage (CAES), as a new storage technology, is mentioned by Nicolosi (2010), given that most of the sites in Europe where hydro-pumped-storage is possible are already explored.

- Integration of electricity markets, including balancing and ancillary service markets, allowing generation optimization and increasing RES-E utilization.

- Rising grid flexibility through the reinforcement of transmission and distribution lines and cross-border interconnections, increasing security of supply and regional imbalances. Additionally, by extending the transmission grid into zones of high availability of renewable resource, the RES-E potential can then be unleashed and used in other high demand zones (Schaber et al. 2012).
- Flexible and efficient generation mix sustained by high price spikes, recognizing scarcity, and allowing investment cost recovery of conventional power. This is fundamental to guarantee the security of supply in the absence of renewable resource and low RES-E generation.
- Capacity mechanisms might be required to support backup dispatchable generation, allowing for investment cost recovery and providing an incentive for new dispatchable, efficient, and low emission power plants (e.g., combined cycle gas turbines) (Henriot and Glachant 2013).

Moreover, intermittency is a well-known characteristic of RES-E due to its nondispatchability and the variable nature of renewable resources. Moreno and Martínez-Val (2011), for example, identify events in Spain when wind power decreases 10 GW within 24 hours coincident with increasing demand of 16 GW within 8 hours. Furthermore, the increasing deployment of photovoltaic power is changing the daily load profile and increases production prediction errors. The limited storability of electrical energy creates a difficulty in balancing these events. An electrical system with high shares of RES-E, as described earlier, needs to be flexible to guarantee a determined reliability level* of supply with or without renewable resources available. Therefore, a short notice electricity supply source is required when RES-E suddenly fails. These supply sources consist of the so-called backup power and have to be adequately compensated, guaranteeing not only their marginal costs but also fixed and investment costs, through adequate scarcity price signals or capacity mechanisms (Henriot and Glachant 2013). This backup power can at first be provided by standby power plants such as

- Power storage—hydro-pump-storage, CAES
- Dispatchable renewables   hydro-dams, biomass
- Thermal power—combined cycle gas turbines (CCGT), coal, and nuclear

Given the GHG emission reduction targets and the flexibility required, it is fundamental to prioritize the development of power storage and

---

* The reliability of a transmission system can be measured by a number of indicators: loss of load expectation, loss of load events, loss of expected energy, expected energy unserved, and value of lost load (Newbery 2015).

dispatchable renewables. Hydro-pump-storage is beyond doubt the main storage technology available, which is capable of storing the large amounts of energy required nowadays. Nevertheless, given the limited sites available to further develop this type of power facilities, incentives should be provided for the research and development of new storage technologies, such as batteries or CAES (Benatia et al. 2013). In the thermal power category, CCGTs are by far the most flexible and efficient (Moreno and Martínez-Val 2011). However, due to the low amount of residual load, hence diminished load factors, their financial sustainability needs to be considered, either through adequate scarcity price signals or capacity mechanisms.

With the implementation of capacity mechanisms, power plants capable of dispatch control are remunerated for providing a power capacity guarantee. This capacity guarantee might be subject to regular testing if the plant is not operated for some time. Capacity mechanisms can be applied in various forms, of which capacity payments, strategic reserves, and capacity markets are the most common (Meulman and Méray 2012). The idea of capacity mechanisms is not undisputed, as it is seen to introduce an additional subsidy and is subject to over-procurement (Hildmann et al. 2015; Newbery 2015). In fact, some authors defend that energy-only markets are able to provide adequate price signals if combined with other measures, such as adequate remuneration of security services, reinforcing transmission grids and cross-border interconnections, and demand response implementation, among others (Henriot and Glachant 2013; Newbery 2015).

Backup power can also be provided by a strong and flexible transmission grid and interconnections. This is a more suitable alternative, compared to a massive supply infrastructure built merely for backup and hard to be financially justified. Besides, reinforcing transmission grids also allows the optimization of other existing production infrastructure, including baseload plants, such as nuclear and coal power plants (Schaber et al. 2012). With a strong transmission grid, surplus amounts of RES-E can be transported to other load centers without grid congestions and the need to proceed with curtailments, thus optimizing RES-E production. This occurs when there is high availability of renewable resource and the RES-E installed capacity is able to produce more electrical energy than the amount demanded. In the absence of adequate transmission grid capacity, the surplus of electrical energy does not have a path to flow and the lines become constrained, leading to selective curtailment of RES-E and inefficiencies. Cross-border interconnections can thenceforth facilitate the trade of these surplus amounts and at the same time provide geographical dispersion and diversification of the generation mix available, improving security of supply and replacing the need for standby generation.

Flexibility of the electricity system can similarly be attained by adequate reactions on the demand side. Demand response or demand-side management is the concept involving consumers responding to short-term price signals and adjusting demand accordingly. Consumers would be able to decrease demand, if adequate incentives are provided, by transferring some loads to lower price periods of the day, including in the future the well-known electric vehicle charging (Benatia et al. 2013). These price signals would be part of smart grid information, to which each consumer would have access through the installed smart-meter. Rising demand elasticity would mitigate the missing money problem and help in balancing supply and demand (Newbery 2015).

No unique answer can be found to the challenge of RES-E market integration, rather a mix of well-adjusted actions should be taken from backup power with storage and thermal, to reinforce transmission grid and demand response—all can play a part in the future electricity system, desired to be reliable and sustainable.

6.5 RES-E Optimization through Market Integration

Market integration in wholesale electricity trading has been intensively pursued by the European policy pursuing the vision of a single energy market since the 1990s. Policy makers have been encouraged by the pursuit of economic efficiency and greater competition, to reinforce interconnectors and harmonize trading rules, given the emergence of substantial amounts of intermittent renewable generation.

A high level of RES-E generation can create transmission grid congestion, thus reinforcing the transmission grid and cross-border interconnections is paramount in RES-E market integration and in regional electricity market integration. As stated in the previous section, this is one of the fundamental actions to be taken to achieve an efficient electricity market.

Cross-border interconnections present numerous advantages, such as production optimization, increasing opportunities for operation with renewable energies, the promotion of competition, and the improvement of supply security by providing backup supply. Yet, the existing limited capacity has to be managed efficiently, allowing for cross border trading. The cross-border interconnection management made through implicit auctions, the market splitting/coupling mechanisms, allows the coordination of different price areas, increasing overall welfare in the electricity markets (Jacottet 2012). Weber et al. (2010) clarify the difference between market splitting and market coupling: in market splitting a single power exchange operates several electricity bidding areas, whilst under market coupling multiple power exchanges cooperate to manage different electricity bidding areas.

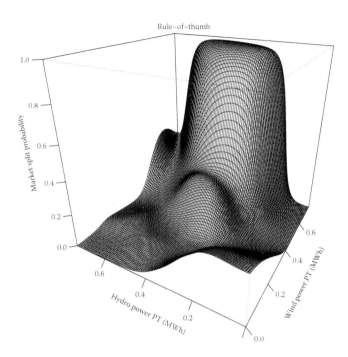

FIGURE 6.10
Predicted probability response to wind and hydro power generation shares in Portugal.

Extensive research on modeling electricity market integration can be found in the literature, expressed both in terms of price convergence and the dynamics of shock transmissions. Price convergence has been modeled by estimating the probability of market splitting between electricity bidding areas. In Figueiredo et al. (2015a,b), nonparametric models were used to estimate market splitting probabilities unveiling its behavior and determinants associated with RES-E. It was shown that different dimensions of electrical systems play a role in the behavior of the electricity market splitting. For example, as shown in Figure 6.10, in Portugal, when we are in the presence of simultaneous high generation of wind and hydropower, market-splitting probability in Iberia increases. Low marginal cost generation is demonstrated to affect market splitting, and therefore cross-border congestion. This is true even considering the high level of cross-border interconnections between Iberian countries.[*]

Coordination between the development of RES-E and reinforcements of transmission grid, including cross-border interconnections, should exist in

---

[*] The current cross-border interconnection capacity is 3000 MW representing 32% of the smallest bidding area peak demand.

the European energy policy. This would allow price convergence between bidding areas to be within reasonable levels, fostering market integration.

6.6 Final Remarks

As we mentioned in the previous sections, the extensive deployment of RES-E in some European electricity markets creates demanding challenges to the electricity sector. RES-E development aims to improve energy security, decrease the dependency on fossil fuels, and reduce greenhouse gas emissions. With the targets set for 2030 by the EU, establishing a RES-E share increase to 45%, RES-E is required to further grow in the electricity system. Given the "merit-order effect," where the low marginal cost RES-E displaces the aggregated supply bid curve to the right, the available residual load decreases dramatically for technologies with higher marginal costs. Additionally, spot electricity prices also decrease and the market fails to provide correct signals to sustain adequate generation capacity, the "missing money problem."

Moreover, RES-E integration into the electricity market requires market adjustments in order to overcome the identified failures. The *melting-pot* and *salad-bowl* express the two alternative routes for policy makers; however, one thing we can ascertain is that flexibility of the electricity system is fundamental to obtain an efficient electricity market. This flexibility can be obtained through a number of strategies, of which regional market integration and demand response seem to be unanimous throughout the literature.

Policy makers pursue regional market integration because it is believed that it will lead to economic efficiency and greater competition, benefiting from cross-border interconnections and trade. It provides the desired electricity system flexibility for RES-E market integration and improves security of supply. Nevertheless, congestion of cross-border interconnections, thus electricity price divergence between bidding areas, is demonstrated to occur with high low marginal cost generation, and consequently reinforcing the transmission grid and cross-border interconnections is vital. Transmission grid and cross-border interconnection expansions should be coordinated with RES-E deployment in order to contribute to electricity price convergence.

Albeit recognizing some factors that influence the deployment of renewables, there is no defined formula to facilitate the integration of high levels of RES-E into the electricity system. Policy makers and stakeholders, in general, have to consider all available strategies and tailor the best possible path, bearing in mind that interactions between regions exist and that the objective is common: to obtain a competitive, reliable, and sustainable electricity system.

Acknowledgments

This work was supported by the *Fundação para a Ciência e a Tecnologia* (FCT) under project grant UID/MULTI/00308/2013.

References

Amorim, F., Vasconcelos, J., Abreu, I.C., da Silva, P.P., Martins, V. 2013. How much room for a competitive electricity generation market in Portugal? *Renewable and Sustainable Energy Reviews* 18: 103–118.

Batlle, C., Pérez-Arriaga, I.J.J., Zambrano-Barragán, P. 2012. Regulatory design for RES-E support mechanisms: Learning curves, market structure, and burden-sharing. *Energy Policy* 41: 212–220.

Benatia, D., Johnstone, N., Haščič, I. 2013. Effectiveness of policies and strategies to increase the capacity utilisation of intermittent renewable power plants. OECD Environment Working Papers, OECD Publishing, Paris no. 57, pp. 1–49.

BP. 2015. Statistical Review of World Energy 2015. http://www.bp.com/en/global/corporate/energy-economics/statistical-review-of-world-energy.html. Accessed March 31, 2016.

Brundtland, G.H. 1987. Our common future: Report of the World Commission on Environment and Development. *Medicine, Conflict and Survival* 4(1): 300.

Cramton, P., Stoft, S. April 2006. The convergence of market designs for adequate generating capacity. A White Paper for the Electricity Oversight Board. https://works.bepress.com/cramton/34/. Accessed April 14, 2016.

Danish Energy Authority. January 2007. A visionary Danish energy policy. Report, p. 30. http://www.ens.dk/sites/ens.dk/files/dokumenter/publikationer/downloads/engelsk_endelig_udgave_visionaer_energipolitika4.pdf. Retrieved on February 27, 2015.

Diekmann, J., Kemfert, C., Neuhoff, K. 2012. The proposed adjustment of Germany's renewable energy law: A critical assessment. *DIW Economic Bulletin* 2: 3–9.

Edenhofer, O., Hirth, L., Knopf, B., Pahle, M., Schlömer, S., Schmid, E., Ueckerdt, F. 2013. On the economics of renewable energy sources. *Energy Economics* 40: 12–23.

Edenhofer, O., Pichs Madruga, R., Sokona, Y., United Nations Environment Programme, World Meteorological Organization, Intergovernmental Panel on Climate Change, Working Group III, Potsdam-Institut für Klimafolgenforschung. 2012. Renewable energy sources and climate change mitigation: Special report of the Intergovernmental Panel on Climate Change. Cambridge University Press, 2011. http://www.cambridge.org/us/academic/subjects/earth-and-environmental-science/climatology-and-climate-change/renewable-energy-sources-and-climate-change-mitigation-special-report-intergovernmental-panel-climate-change.

EPEX, APX-endex, and BelPEX. June 2010. CWE market coupling algorithm, pp. 1–19. http://static.epexspot.com/document/20015/COSMOS_public_description. pdf. Accessed November 11, 2012.

European Union. 2009a. Directive 2009/28/EC of the European Parliament and of the Council of 23 April 2009 on the promotion of the use of energy from renewable sources and amending and subsequently repealing Directives 2001/77/EC and 2003/30/EC. Official Journal of the European Union L140 (5-6-2009), pp. 16–62.

European Union. 2009b. Directive 2009/72/EC of the European Parliament and of the Council of 13 July 2009 concerning common rules for the internal market in electricity and repealing Directive 2003/54/EC. Official Journal of the European Union L 211 (14/08/2009), pp. 55–93.

Eydeland, A., Wolyniec, K. 2003. *Energy and Power Risk Management*. John Wiley & Sons, Inc. http://eu.wiley.com/WileyCDA/WileyTitle/productCd-0471104000. html.

Felder, F.A. 2011. Examining electricity price suppression due to renewable resources and other grid investments. *The Electricity Journal* 24(4): 34–46.

Figueiredo, N.C., da Silva, P.P., Cerqueira, P.A. 2015a. Wind generation influence on market splitting: The Iberian spot electricity market. In *2015 12th International Conference on the European Energy Market (EEM)*. IEEE, pp. 1–5. http://ieeexplore.ieee.org/lpdocs/epic03/wrapper.htm?arnumber= 7216649.

Figueiredo, N.C., da Silva, P.P., Cerqueira, P.A. 2015b. Evaluating the market splitting determinants: Evidence from the Iberian spot electricity prices. *Energy Policy* 85: 218–234.

Freitas, C., da Silva, P.P. 2015. European Union Emissions Trading Scheme impact on the Spanish electricity price during phase II and phase III implementation. *Utilities Policy* 33: 54–62.

Gelabert, L., Labandeira, X., Linares, P. 2011. An ex-post analysis of the effect of renewables and cogeneration on Spanish electricity prices. *Energy Economics* 33: S59–S65.

Henriot, A., Glachant, J.M. 2013. Melting-pots and salad bowls: The current debate on electricity market design for integration of intermittent RES. *Utilities Policy* 27: 57–64.

Hildmann, M., Ulbig, A., Andersson, G. July 2015. Revisiting the merit-order effect of renewable energy sources, pp. 1–13. IEEE Power & Energy Society General Meeting, pp. 1-1. http://arxiv.org/abs/1307.0444.

Hirschhausen, C. 2014. The German 'Energiewende'—An introduction. *Economics of Energy & Environmental Policy* 3(2): 1–12.

Jacottet, A. June 2012. Cross-border electricity interconnections for a well-functioning EU internal electricity market, pp. 1–17. https://www.oxfordenergy.org/ publications/cross-border-electricity-interconnections-for-a-well-functioning-eu-internal-electricity-market-2/. Accessed July 3, 2012.

Jager, D., Klessmann, C., Stricker, E., Winkel, T., de Visser, E., Koper, M., Ragwitz, M., Held, A. 2011. Financing renewable energy in the European energy market. Report, Ecofys, no. PECPNL084659, pp. 1–264. http://ec.europa.eu/energy/ renewables/studies/doc/renewables/2011_financing_renewable.pdf. Accessed May 9, 2012.

Klessmann, C., Nabe, C., Burges, K. 2008. Pros and cons of exposing renewables to electricity market risks—A comparison of the market integration approaches in Germany, Spain, and the UK. *Energy Policy* 36(10): 3646–3661.

Lund, H. 2010. The implementation of renewable energy systems. Lessons learned from the Danish case. *Energy* 35(10): 4003–4009.

Lund, H., Hvelplund, F., Østergaard, P.A., Möller, B., Vad Mathiesen, B., Karnøe, P., Andersen, A.N. et al. 2013. System and market integration of wind power in Denmark. *Energy Strategy Reviews* 1: 143–156.

Lynch, M.Á., Tol, R.S.J., O'Malley, M.J. 2012. Optimal interconnection and renewable targets for north-west Europe. *Energy Policy* 51: 605–617.

Mauritzen, J. 2010. What happens when it's windy in Denmark? An empirical analysis of wind power on price volatility in the Nordic electricity market. *SSRN Electronic Journal* 1–29.

Meulman, L., Méray, N. 2012. Capacity mechanisms in northwest Europe. Clingendael International Energy Programme, pp. 1–62. http://clingendael.info/publications/2012/20121200_capacity_mechanisms_in_northwest_europe_december.pdf.

Meyer, N.I. 2003. European schemes for promoting renewables in liberalised markets. *Energy Policy* 31(7): 665–676.

MIT Energy Initiative. 2011. Managing large-scale penetration of intermittent renewables. Report, 115, pp. 1–240. http://energy.mit.edu/publication/managing-large-scale-penetration-of-intermittent-renewables/. Accessed March 2, 2015.

Moreno, B., Pereira da Silva, P., 2016. How do Spanish polluting sectors' stock market returns react to European Union allowances prices? A panel data approach. *Energy*, 103: 240–250.

Moreno, F., Martínez-Val, J.M. 2011. Collateral effects of renewable energies deployment in Spain: Impact on thermal power plants performance and management. *Energy Policy* 39(10): 6561–6574.

Nelson, T., Reid, C., McNeill, J. 2015. Energy-only markets and renewable energy targets: Complementary policy or policy collision? *Economic Analysis and Policy* 46: 25–42.

Newbery, D. 2015. Missing money and missing markets: Reliability, capacity auctions and interconnectors. *Energy Policy* 94: 401–410.

Nicolosi, M. 2010. Wind power integration and power system flexibility—An empirical analysis of extreme events in Germany under the new negative price regime. *Energy Policy* 38(11): 7257–7268.

Pereira da Silva, P., 2007. *O Sector Da Energia Eléctrica Na União Europeia: Evolução E Perspectivas.* Coimbra University Press.

Pereira da Silva, P., Moreno, B., Figueiredo, N. 2016. Firm-specific impacts of CO_2 prices on the stock market value of the Spanish power industry. *Energy Policy* 103: 240–250.

Pereira da Silva, P., Moreno, B., Fonseca, A.R. 2015. Towards an auction system in the allocation of EU emission rights: Its effect on firms' stock market returns. In *Assessment Methodologies—Energy, Mobility and Other Real World Applications*, eds. Godinho, P. and Dias, J., pp. 33–64. Coimbra, Portugal: Coimbra University Press.

Rivier Abbad, J. 2010. Electricity market participation of wind farms: The success story of the Spanish pragmatism. *Energy Policy* 38(7): 3174–3179.

Schaber, K., Steinke, F., Hamacher, T. 2012. Transmission grid extensions for the integration of variable renewable energies in Europe: Who benefits where? *Energy Policy* 43: 123–135.

Sensfuß, F., Ragwitz, M., Genoese, M. 2008. The merit-order effect: A detailed analysis of the price effect of renewable electricity generation on spot market prices in Germany. *Energy Policy* 36(8): 3076–3084.

Traber, T., Kemfert, C. 2011. Gone with the wind?—Electricity market prices and incentives to invest in thermal power plants under increasing wind energy supply. *Energy Economics* 33(2): 249–256.

Weber, A., Graeber, D., Semmig, A. 2010. Market coupling and the CWE project. *Zeitschrift Für Energiewirtschaft* 34(4): 303–309.

Würzburg, K., Labandeira, X., Linares, P. 2013. Renewable generation and electricity prices: Taking stock and new evidence for Germany and Austria. *Energy Economics* 40: S159–S171.

7

Optimal Scheduling of a Microgrid under Uncertainty Condition

Gabriella Ferruzzi and Giorgio Graditi

CONTENTS

7.1 Introduction

Due to the increase of the energy demand, to the obsolescence of the HV grid, to the improving of the sensitivity to environmental issue, micro-grids (MGs) and smart-grids (SGs) can become a real opportunity of success.

These are conceived as electric grids in low voltage (LV) and medium voltage (MV), respectively, able to deliver electricity in a controlled smart way from points of generation to consumers. Through the two-way flow of information between suppliers and consumers, the new grids encourage users' participation in energy saving and their cooperation through the demand–response mechanism.

Several investigators have analyzed the role played by MGs/SGs in terms of energy price reduction or reliability system improvement, as well as their impact on the operating costs reduction or on environmental aspects.

In this chapter, a risk management model for the day-ahead energy market is proposed to determine optimal economic choices for the management of an MG that works under uncertainty conditions.

The chapter is divided into four sections. In the first, the MG concept is described; in the second, the market structure is analyzed; in the third, several methodologies to evaluate the uncertainties are discussed; and in the last section, a case study is presented.

7.2 Microgrids

7.2.1 Definitions

The power grid consists of various electrical components and at multiple levels: transmission in high voltage (HV), distribution in medium voltage (MV), and distribution in low voltage (LV).

In this framework, the microgrids (MGs) are classified as a distribution grid with distributed energy resources (microturbines, fuel cells, photovoltaics—PV, etc.) and storage devices (flywheels, energy capacitors, and batteries), usually in LV, able to provide services in both autonomous (island) and grid-connected modes (Figure 7.1).

Different components, designs, and rules are defined by the manager of an MG, who aggregates the capacity of different components and buys or sells, for each hour, power from/to the grid with higher-level voltage (Lassater, 2001; Schwaegerl et al., 2009b; Del Carpio et al., 2010; El-hawary, 2014; Graditi et al., 2016a).

From the grid's point of view, an MG is as a controlled entity within the power system that can operate as a single aggregated load; from a customer's point of view, MGs not only provide their thermal and electricity needs, but in addition, enhance local reliability, reduce emissions, improve power quality, and can contribute to the accommodation of electric vehicles and storages (Figure 7.2). In this framework, the concept of control and management assumes a key role (Favuzza et al., 2006; Schwaegerl et al., 2009a,b, 2010).

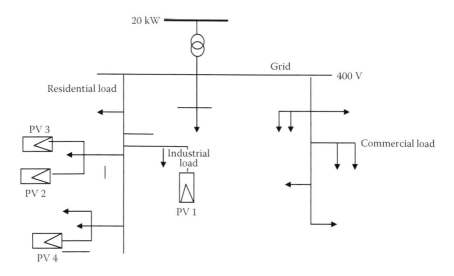

FIGURE 7.1
Microgrid scheme.

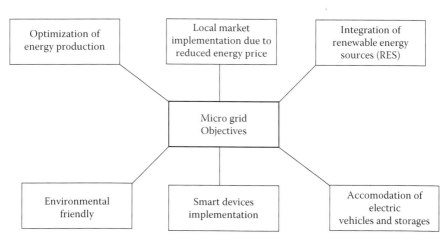

FIGURE 7.2
Microgrid advantages.

7.2.2 Structures of Control

There is no general structure of MG control architecture, since the configuration depends on the type of MG or the existing infrastructure, but, independent of the architecture, the hierarchic scheme comprises the three following levels (Schwaegerl et al., 2009b; Mahmoud et al., 2014):

- Local microsource controllers (MCs) and load controllers (LCs)
- Microgrid central controller (MGCC)
- Central autonomous management controller (CAMC)

With regard to the function of the responsibilities assumed by the different control levels, two different structures of control are recognized: centralized and decentralized.

In centralized control, the main responsibility for the maximization of the MG value is the microgrid central controller (MGCC), which provides the main interface between the MG and other actors and can assume different roles, ranging from the main responsibility for the maximization of the MG value to the simple coordination of the local MCs. Finally, MGCC sends dispatch signals to both the MCs and LCs (Tsikalakis and Hatziargyriou, 2008; Schwaegerl et al., 2009b; Kaur et al., 2016).

In a fully decentralized approach, the main responsibility is given to the MCs that compete to optimize their production, taking into account current market prices.

This is based on the multiagent* system (MAS) theory. The core idea is that an autonomous control process is assumed by each controllable element. The MAS theory describes the relationship between the agents and the organization of the whole system, although the agent does not know the status of the whole system. This approach is suitable in cases of different ownership of distributed energy resources (DERs), where several decisions need to be taken locally, making centralized control very difficult (Dimeas and Hatziargyriou, 2004; Schwaegerl et al., 2009b; Cai and Mitra, 2010; Luu, 2014; Kantamneni et al., 2015).

---

* There is no formal definition of an agent, but in the literature (Alfredo et al., 2012; Alessandrini et al., 2014) the following basic characteristics are provided:
- An agent can be a physical entity that acts in the environment or a virtual one, that is, with no physical existence.
- An agent is capable of acting in the environment, that is, the agent changes its environment by its actions.
- Agents communicate with each other, and this could be regarded as part of their capability for acting in the environment.
- Agents have a certain level of autonomy, which means that they can take decisions without a central controller or commander. To achieve this, they are driven by a set of tendencies.

7.2.3 Microgrid Ownership and Business Model

There is a very strong relation between the MG owner and the business model used.

In order to identify the impact of asset ownership on financial interactions among various MG stakeholders, three main business models are introduced: the first one is the distribution system operator (DSO) monopoly model, the second one is the prosumer consortium, and the last is the free market model.

The main difference lies in the ownership of DERs.

In the DSO monopoly model, in particular, the distribution system operator has ownership of the DERs as well as the electric grid; in the second approach, a manager of the grid, that is, "prosumer," manages and coordinates the different agents that belong to the MG to maximize the revenues of the system; whereas, in the last approach, each agent acts to maximize own benefit (Schwaegerl et al., 2009a,b).

7.2.3.1 The DSO Monopoly Model

A DSO monopoly type of MG has very probably evolved from a nonliberalized power industry because the DSO not only owns the distribution grid but also assumes the retailer function of selling electricity to end consumers.

In general, a DSO monopoly microgrid is likely to be built upon a technically challenged distribution grid with aging, maintenance, and/or supply quality problems.

In a DSO monopoly MG, DERs tend to be larger, and storage units tend to be located at substations (Schwaegerl et al., 2009b).

7.2.3.2 The Prosumer Consortium Model

In this case, the consortium works to minimize the total costs of the system reducing the internal energy consumption, or maximize the revenues derived by the electricity export.

This type of MG may find considerable barriers set by the DSO, as by nature the consortium tends to minimize the use of distribution grid and may neglect all network constraints during the design of the microgrid. In a prosumer consortium microgrid, DERs tend to be smaller, and storage tends to be small and dispersed (Schwaegerl et al., 2009b).

7.2.3.3 The Free Market Model

In this case, the MGCC will behave as an energy retailer that is simultaneously responsible for local balance, import and export control, technical performance maintenance, and emission level monitoring. In a free market MG, DER and storage can vary in forms, sizes, and locations (Schwaegerl et al., 2009b).

7.3 Deliberalized Energy Market Structure

The MGs can take part in the energy market acting as individual market players that provide/buy energy to/from the system.

Two different trading markets* are usually available to facilitate energy commerce between producers and consumers: futures and pool markets.

In the first, long-term contracts are exchanged between producers and consumers to limit the energy price volatility. In the second one, power/energy quantities are exchanged between producers and consumers in order to minimize the energy price (Rossi, 2007; Cai and Mitra, 2010).

7.3.1 Agents in the Deregulated Electricity Market

The market includes different agents as consumers, retailers, producers, the market operator (MO), the independent system operator (ISO), and the regulator.

The first three agents participate in the energy market as "profit agents," that is, they want to maximize their profit or minimize their cost, whereas the last three agents supervise, manage, and control so that the system works in the right way.

Consumers are the end users of the electricity and may purchase energy in the pool or be supplied by retailers. A consumer aims to minimize their procurement cost or to maximize the utility they obtain from electricity usage. Retailers provide electricity to the end consumers and aim to maximize the profit they obtain from selling to their customers. Producers are in charge of the electricity generation and sell electric energy either to the electricity markets (pool and futures market) or directly to the consumers and the retailers.

A market operator (MO) is a no-profit entity responsible for the economic management of the marketplace as a whole. In addition, the MO administers the market rules and determines the prices and quantities of energy traded in the market (Rossi, 2007).

An independent system operator (ISO) is a no-profit entity in charge of the technical management of the electric energy system pertaining to the electricity market. The ISO provides equal access to the grid to all consumers, retailers, and producers. A market regulator (MR) is a government-independent entity whose function is to oversee the market and to ensure its competitive and adequate functioning.

Additionally, the MR promotes and enforces orders and regulations (Schwaegerl et al., 2010).

---

* There also exists the possibility of bilateral contracts between suppliers and consumers defined outside an organized market place.

7.3.2 The Pool

The pool is a marketplace where the energy is traded and typically includes (Figure 7.3)

1. A day-ahead market
2. Several adjustment markets
3. Balancing markets

In the pool, producers submit production offers while consumers and retailers submit consumption bids to the day-ahead, adjustment, and balancing markets, and in turn, the MO clears these markets and determines prices and traded quantities.

The energy traded in the pool is mostly negotiated in the day-ahead market, while adjustment markets are used to make adjustments to the output of the day-ahead market.

In the day-ahead and adjustment markets, producers submit energy blocks and their corresponding minimum selling prices for every hour of the market horizon and every production unit. At the same time, retailers and consumers submit energy blocks and their corresponding maximum buying prices for every hour of the market horizon (Rossi, 2007).

The MO collects purchase bids and sale offers and clears the market (both day-ahead and adjustment) using a market-clearing procedure.

A market-clearing procedure results in market-clearing prices, as well as production and consumption schedules. If the transmission grid is not considered in the market-clearing procedure, the resulting market-clearing price is identical for all market agents.

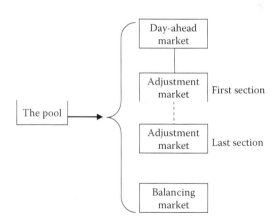

FIGURE 7.3
The pool structure.

On the other hand, if the transmission network is taken into account for clearing the market, instead of a single market-clearing price, a locational marginal price (LMP) is associated with each node of the power system.

The balancing market, cleared on an hourly basis (or several times within each hour) through an auction, provides energy to cover both generation excess and deficit and constitutes the last market prior to power delivery to balance production and consumption.

Producers/consumers submit balancing offers that are accepted by the MO on an increasing price basis until balance is guaranteed in the case of deficit of generation.

Alternatively, for the case of excess of generation, offers to reduce production are accepted on a decreasing price basis until balance is ensured.

Producers participate providing balancing (up and down) energy, while nondispatchable producers and consumers use this market to self-balance their energy productions and consumptions, respectively, to those values agreed upon in previous pool markets. Retailers that behave as consumers are not represented in this figure for the sake of simplicity. The balancing market ensures a balanced system operation (Rossi, 2007).

7.4 Uncertainties Evaluation

Forecasting is an important tool and a crucial factor, especially in the deregulated energy market where a decision maker has the need of accurate forecasts of future demands, energy prices, and also fossil fuel to maximize revenues.

Depending on the time horizon and the operating decisions that need to be made, different forecasts are needed: short-term, medium-term, and long-term forecasting.

In general, long-term forecasting is needed for system planning and economic analyses; medium-term forecasting is needed for maintenance of the system; finally, short-term forecasting is needed for the day-to-day operation of the system (Conejo et al., 2010; Silva et al., 2011).

In this work, short-term forecasting (for electricity price, load, and intermittence power production) is treated in depth because the accuracy of these allows significant saving operating costs and an enhanced system reliability.

7.4.1 Uncertainty Factors

7.4.1.1 Load

Electricity demand forecasts are extremely important for energy suppliers and other participants in electric energy generation, transmission, distribution, and markets. Various techniques and models have been developed for the forecasting of electrical load with varying degrees of success.

Load series exhibits several levels of seasonality; the prediction does not depend only on the previous hour load, but also on the load of the same hour on previous day, and same denominations in the previous week (Bunn, 2000; Hesham et al., 2002; Feinberg and Genethliou, 2005; Khan et al., 2006; Kyriakides and Polycarpou, 2007).

According to the forecasting horizon, load forecasting can be broadly divided into three categories: short-term forecasts, which are usually from 1 hour to 1 week; medium forecasts, which are usually from a week to a year; and long-term forecasts, which are longer than a year.

7.4.1.2 Electricity Prices

Electricity price forecasting is characterized by time-of-the-day effect, multiple seasonality, high volatility, and nonstationarity mean and variance.

Various techniques and models have been developed for the forecasting of electrical price with varying degrees of success. According to the forecasting horizon, price forecasting can be broadly divided into three categories: short-term forecast that covers time intervals ranging from less than 1 hour to a few hours; medium forecast that covers several hours to a few days ahead; and long-term forecast that covers seasonal to annual horizons (Bunn, 2000; Khan et al., 2006; Hu et al., 2009; Jain et al., 2013; Weron, 2014).

7.4.1.3 Renewable Power Production

In the relevant literature, various forecasting methods have been proposed to estimate the expected power generated from a renewable energy source, which essentially differ in the type of the information characterizing the predicted output and in the time horizon of their application. The methodologies applied combine multidisciplinary fields and areas such as meteorology, statistics, physical modeling, and computational intelligence. According to the forecasting horizon, forecasting can be broadly divided into three categories: short-term forecast that covers time intervals ranging from less than 1 hour to few hours; medium forecast that covers several hours to few days ahead; long-term forecast that covers seasonal to annual horizons (Mureddu et al., 2015).

7.4.2 Forecasting Techniques

A large number of methods and techniques have been developed and various approaches have been introduced. They can be grouped, usually, into two main classes: classical/conventional methods and computational intelligence-based techniques. The first category includes methods such as time series models, regression models, and Kalman filtering-based techniques. Computational intelligence-based techniques include expert systems,

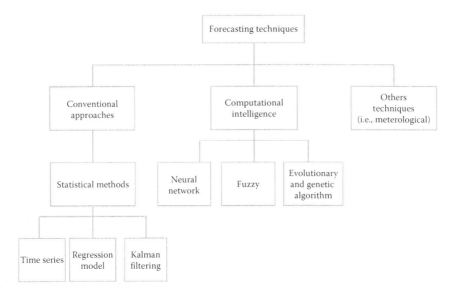

FIGURE 7.4
Forecasting techniques classification.

artificial neural networks, fuzzy inference and fuzzy–neural models, and evolutionary programming.

In this work, another class of techniques will also be considered to predict the amount of renewable energy production using weather forecasts as a general input to the various forecasting functions.

This section offers an overview of the various approaches (Bunn, 2000; Hesham et al., 2002; Feinberg and Genethliou, 2005; Khan et al., 2006; Kyriakides and Polycarpou, 2007; Hu et al., 2009; Jain et al., 2013; Weron, 2014; Graditi et al., 2016b), as shown in Figure 7.4.

7.4.2.1 Conventional Approach

7.4.2.1.1 Time Series Models

Time series analysis is a method of forecasting that focuses on the past behavior of the dependent variable. Using the time series approach, a model is first developed based on the previous data, and then future variable is predicted based on the model. The basic assumption of stationarity on the error terms includes zero mean and constant variance.

These techniques assume that the data follow a certain stationary pattern that depends on autocorrelation; trends in the data; and daily, weekly, and seasonal variations.

Time series models appear in the literature in different forms such as Box–Jenkins, time series, stochastic models, autoregressive moving average (ARMA), autoregressive integrated moving average (ARIMA), auto regressive moving average with exogenous variables (ARMAX), autoregressive integrated moving average with exogenous variables (ARIMAX), and state-space models.

The basic idea in time series prediction is to model the variable of interest as the sum of two terms, $z(t) = yp(t) + y(t)$, where $yp(t)$ is the contribution to the system that depends on the time of day, while $y(t)$ is a residual term that models the deviation of the weather pattern.

The residual term may be modeled by:

$$y(t) = \sum_{i=1}^{n} a_i y(t-i) + \sum_{k=1}^{n_u} \sum_{j_k=0}^{m_k} b_{j_k} u_k(t-j_k) + \sum_{h=1}^{H} c_h w(t-h) \tag{7.1}$$

where

$u_k(t)$, $k=1,2,\ldots,n_u$ represent the inputs

$w(t)$ is a zero-mean white random process that represents uncertain effects on load demand and random load behavior

The goal is to identify the parameters a_i, b_{jk}, and c_h and the integers n, n_u, m_k, and H by fitting the model using historical data.

In general, time series methods give satisfactory results if there is no change in the variables that affect load demand (i.e., environmental variables).

7.4.2.1.2 Moving Average (MA) Models

The moving average (MA) model is a common approach for modeling univariate time series. With the moving average modeling technique, the current value of time series $Y(t)$ is expressed in terms of a linear combination of current and previous values of white noise series.

Mathematically it is expressed as:

$$Y(t) = \mu + \varepsilon_t + \varphi_1 \varepsilon_{t-1} + \cdots + \varphi_q \varepsilon_{t-q} \tag{7.2}$$

where

μ is the mean of the series

$\varphi_1, \ldots, \varphi_q$ are the parameters of the model

$\varepsilon_{t-1}, \ldots, \varepsilon_{t-q}$ are white noise error terms

The value of q is called the order of the MA model.

The backshift operator on white noise modified is:

$$Y(t) = \varphi(B) a(t) \tag{7.3}$$

where

$$\varphi(B) = 1 - \varphi_1 B - \varphi_2 B^2 - \cdots - \varphi_q B^q \tag{7.4}$$

This can be equivalently written in terms of the backshift operator B as:

$$Y(t) = \mu + \left(1 + \varphi_1 B + \cdots + \varphi_q B^q\right)\varepsilon_t \qquad (7.5)$$

7.4.2.1.3 Autoregressive (ARMA)

The autoregressive moving average model takes into account the random nature and time correlations of the phenomenon under study.

In this model, the current values of time series $Y(t)$ express linearly in terms of the previous period $(y(t-1), y(t-2), \ldots)$ and current and previous values of white noise $(a(t), a(t-1), a(t-2), \ldots)$.

The mathematical model is as follows:

$$Y(t) = \Phi_1 y(t-1) + \cdots + \Phi_p y(p-t) + \cdots + a(t-1) + \cdots + \Phi_q a(t-q) \qquad (7.6)$$

In the ARMA(p, q) model, the current value of the price y_t is expressed linearly in terms of its p past values (autoregressive part) and in terms of q previous values of the noise (moving average part): $\varphi(B)X_t = \theta(B)\varepsilon_t$.

Here, B is the backward shift operator, $\theta(B)$ is a shorthand notation for $\theta(B) = 1 + \theta_1 B + \cdots + \theta_q B_q$, where $\varphi_1, \ldots, \varphi_p$ and $\theta_1, \ldots, \theta_q$ are the coefficients of autoregressive and moving average polynomials, respectively.

Finally, ε_t is the noise (or white noise) with zero mean and finite variance, which is often denoted by White Noise $(0, \sigma^2)$. For $q = 0$, we obtain the well-known AutoRegressive AR(p) model, and for $p = 0$, we get the Moving Average MA(q) model.

7.4.2.1.4 ARIMA Model

If a process is nonstationary, it should be transformed into a stationary process.

If a process is nonstationary, it should be transformed into a stationary process by introducing ∇ operator.

The autoregressive integrated moving average (ARIMA) or Box–Jenkins model has three types of parameters: the autoregressive parameters $(\varphi_1, \ldots, \varphi_p)$, the number of differencing passes at lag-1 (d), and the moving average parameters $(\theta_1, \ldots, \theta_q)$.

A series that needs to be differenced d times at lag-1 and afterward has orders p and q of the AR and MA components, respectively, is denoted by ARIMA(p, d, q) and can be written conveniently as $\varphi(B)\nabla d X_t = \theta(B)\varepsilon_t$, where $\nabla X_t \equiv (1 - B)x_t$ is the lag-1 differencing operator, which is a special case of the more general lag-h differencing operator

$$\nabla h X_t \equiv (1 - Bh)x_t \equiv x_t - x_t - h.$$

Note that ARIMA(p, 0, q) is simply an ARMA(p, q) process.

7.4.2.1.5 Regression Models

The regression-type forecasting model is based on the relationship between several dependent variables and a number of independent variables that are known or estimated.

Although regression-based methods are widely used by electric utilities, they suffer from a number of drawbacks. Due to the nonlinear and complex relationship between the load demand and the influencing factors, it is not simple to develop an accurate model.

One of the main reasons for this drawback is that the model is linearized in order to estimate its coefficients. However, the load patterns are nonlinear and it is not possible to represent the load demand during distinct time periods using a linearized model.

Finally, as with time series methods, regression-based methods may suffer from numerical instability.

The proposed procedure requires few parameters that can be easily calculated from historical data by applying the cross-validation technique.

Multiple regression is based on least squares: the model is fitted such that the sum-of-squares of the differences between observed and predicted values is minimized. In its classical form, multiple regression assumes that the relationship between variables is linear:

$$P_t = BX_t + \varepsilon_t = b_1 X_t^{(1)} + \cdots + b_k X_t^{(k)} + \varepsilon_t \tag{7.7}$$

where
  B is a vector of constant coefficients
  X_t is the vector of regressors
  ε_t is an error term

This model helps in developing a relationship between load, weather condition, day type, consumer class, and so on; it is used for the analysis of measured load as well as price.

7.4.2.1.6 Kalman Filtering Based Techniques

Kalman filtering is based on a particular method of characterizing dynamic systems called state-space representation or state-space model. The Kalman filter is an algorithm for adaptively estimating the state of the model.

The problem formulation of the Kalman filtering approach includes the presence of additive stochastic terms influencing the state and output variables.

7.4.2.2 Computational Intelligence

7.4.2.2.1 Artificial Neural Network (ANN)

The artificial neural networks (ANNs) are highly interconnected, simple processing units designed to model how the human brain performs a particular task. The main idea behind the use of neural networks for forecasting is the assumption that there exists a nonlinear function that relates past values and some external variables to future values of the time series.

During the training process, neurons in the input layer pass the raw information onto the rest of the neurons in the other layers, without any processing. The weights between neurons keep updating according to supervised learning. Based on the measures of minimal error between the output produced and the desired output, the process is repeated until an acceptable error is reached. This training process is called back propagation. After the model acquires the knowledge, new data can be tested for forecasting.

There are three steps that need to be considered in using neural network models for time series prediction: designing the neural network model, training the network, and testing the trained network on a data set that has not been used during the training.

Due to their nonlinear approximation capabilities and the availability of convenient methods for training, artificial neural networks are among the most commonly used methods for electricity load forecasting, especially during the last 10 years.

7.4.2.3 Other Approach: The Weather Predictions

Research in renewable energy forecasting is a multidisciplinary field, since it combines areas such as statistics (Daoud et al., 2012; Bracale et al., 2013; Mureddu et al., 2015), physical modeling, meteorology (Delle Monache et al., 2013; Alessandrini et al., 2014, 2015), and computational intelligence.

The analog method (AM) is shown in the following: it assumes that similar meteorological situations lead to similar local effects. Since the development of numerical weather prediction (NWP) modeling, AM has been used as a statistical adaption of model outputs. For a given target situation forecasting by the NWP model, the general principle of the AM consists in searching for the most similar meteorological situations observed in a historical archive using similar criteria (Alessandrini et al., 2014, 2015).

7.4.2.3.1 Analogs Ensemble (AnEn)

The AnEn is able to estimate the pdf (probability density function) of forecasts solutions by sampling the uncertainty in the analysis and running a number of forecasts from perturbed analyses. The uncertainty of PV energy production is estimated, being the main limiting factor for the participation of an MG in the day-ahead market.

The AnEn methodology is considered to estimate this uncertainty. It uses a single deterministic meteorological forecast and a historical series of past forecasts and associated energy production to generate PV power probabilistic predictions. AnEn selects the historical forecasts most similar to the current prediction and generates probabilistic forecasts of power produced by aggregating the observed historical energy production associated with the selected historical forecasts (Kumar and Kumar, 2011; Alessandrini et al., 2014, 2015). For each forecast lead time and location, the ensemble prediction of solar power is constituted by a set of past production data. These measurements are those concurrent to past deterministic numerical weather prediction forecasts for the same lead time and location, chosen based on their similarity to the current forecast, and in the current application, are represented by the 1 hour average produced solar power.

7.4.2.3.2 Fuzzy Inference and Fuzzy–Neural Models

A relatively new research venture is the combination of fuzzy logic techniques and artificial neural networks to develop forecasting algorithms that merge some of the properties specific to each methodology.

The fuzzy–neural forecasters are typically combined in four different ways:

1. The neural network performs the forecasting and the fuzzy logic system is used to determine the final output.
2. The data are preprocessed using fuzzy logic to remove uncertainties and subsequently a neural network is used to calculate the load estimates.
3. Integrated fuzzy–neural systems where the hidden nodes of the neural network correspond to individual fuzzy rules that are adaptively modified during the training process.
4. Separate neural and fuzzy systems that perform a forecast of different components of the load; these components are then combined at the output to calculate the total load demand.

The combination of fuzzy logic and artificial neural networks creates a hybrid system that is able to combine the advantages of each technique and diminish their disadvantages. The main advantages of the hybrid system are the ability to respond accurately to unexpected changes in the input variables, the ability to learn from experience, and the ability to synthesize new relationships between the load demand and the input variables.

7.4.2.3.3 Evolutionary Programming and Genetic Algorithms

These methods do not get stuck in local minima and can perform well even with noisy data. However, these benefits come at the cost of slow convergence; thus, significant computation periods are needed. One of the applications of evolutionary programming in short-term load forecasting is in connection with time series models.

Then, the evolutionary algorithm is implemented to force the elements of the population of possible solutions to compete with each other and create offspring that approach the optimal solution. The competition for survival is stochastic; the members of the population (parents and offspring) compete with randomly selected individuals based on a "win" criterion. The members of the population are then ranked according to their score and the first half of the population become the parents of the next generation. The process stops once the fitness values of the new generation are not improved significantly.

7.5 Case Study

In this case study, the risk-bidding strategy for the day-ahead energy market is proposed to determine optimal economic choices for the management of a grid-connected residential MG (Kumar and Kumar, 2011; Taheri et al., 2012). It is assumed that the MG consists of different power generation units and traditional power plants, combined heat and power (CHP) generators, PV system, and independent boiler, and that it is controlled and managed by a prosumer. The prosumer participates in the electricity market and needs to determine the optimal bidding (Timmerman and Huitema, 2009; Vogt et al., 2010; Shandurkova et al., 2012; Ferruzzi et al., 2015, 2016; Ottensen et al., 2016).

PV power forecast and the uncertainty associated with its electricity generation have been evaluated through the AnEn approach: the choice of the AnEn approach has been made considering the suitable features of this probabilistic method, such as statistical consistency, reliability, resolution, and skill.

Let Ω_{C} be the set of CHP plants, Ω_{B} be the set of heat production plants, and Ω_{G} be the set of power plants that only produce electricity. $P_{C_{et,j}}$ indicates the power of the jth unit of CHP generation production at the tth hour; $P_{G_{t,j}}$ is the power of the jth unit of only electricity production at the tth hour; $P_{B_{t,j}}$ is the thermal power of the jth heat production at the tth hour; and P_{grid_t} is the power interchange with the MV distribution network at the tth hour. The latter is assumed positive if it is bought from the utility grid and negative if it is sold to the utility grid. Finally, C_{C_j} indicates the power production cost of the jth unit of CHP; C_{g_j} is the power production cost of the jth thermoelectric unit, and C_{g_j} is the thermal production cost of the jth heat production. ρ_t^e is the energy price at the tth hour, which is assumed equal for both buying and selling.

Then, the optimization problem consists of minimizing the following function under a set of technical and operational constraints, as in Ferruzzi et al. (2015, 2016):

$$\sum_{t=1}^{24}\left\{\left[\sum_{j\in\Omega_C}C_{c_j}\left(P_{C_{et,j}}\right)+\sum_{j\in\Omega_B}C_{B_j}\left(P_{B_{t,j}}\right)+\sum_{j\in\Omega_C}C_{G_j}\left(P_{G_{t,j}}\right)+\rho_t^e P_{grid_t}\right]\right.$$
$$\left.+\sum_{j\in\Omega RES}\sum_{p=1}^{n}\xi\rho_t^e\left(x_t-x_t^p\right)\right\} \tag{7.8}$$

where ε is a weight (Mongin, 1997) that takes into account the case in which $\left(x_t - x_t^p\right)$ is positive (overproduction) or negative (underproduction), which is only caused by PV power generation.

The term $\left(x_t - x_t^p\right)$ represents the difference between the expected value of the power produced by the PV plant and the probabilistic value of the analogs.

The introduced weight assumes different values in underproduction and overproduction cases. It is important to factor it into the model because several countries have in place legislation requiring power producers to pay penalties proportional to the errors of the day-ahead energy forecast, which makes the accuracy of such prediction a determining factor for producers to reduce their economic losses.

In Equation 7.9, the hourly production costs can be expressed by the following functional relations:

$$C_{C_j}\left(P_{C_{et,j}}\right) = \alpha_{c_j} P_{C_{et,j}}^2 + \beta_{c_j} P_{C_{et,j}} + \gamma_{c_j}$$
$$C_{G_j}\left(P_{G_{t,j}}\right) = \alpha_{G_j} P_{G_{t,j}}^2 + \beta_{G_j} P_{G_{t,j}} + \gamma_{G_j} \qquad (7.9)$$
$$C_{B_j}\left(P_{B_{t,j}}\right) = \beta_{B_j} P_{B_{t,j}} + \gamma_{B_j}$$

with $\alpha_{c_j}, \beta_{c_j}, \gamma_{c_j}, \alpha_{G_j}, \beta_{G_j}, \gamma_{G_j}, \beta_{B_j}, \gamma_{B_j}$ economic coefficients related to the particular technologies used.

In the following simulation, we assume different power generation units: two traditional power plants, four cogenerators (CHP), and an independent boiler for the generation of thermal energy. Their technical and economic characteristics are given in Table 7.1.

According to Equation (7.2), Figure 7.5 shows the power productions for each technology mentioned related to various energy prices. There is also a 400 kW photovoltaic plant.

The electrical load demand is the aggregate of the loads of six different entities, namely a hotel, a sport center, a hospital, a manufacturing plant, a supermarket, and several offices.

Prosumer, using market prices of electricity, determines the amount of power that the MG should import from the upstream distribution system, optimizing local production or consumption capabilities.

We assume that there are three different prices for each hour (peak, mean, and low price).

Spot prices are obtained using historical data for 12 consecutive months of the Italian electricity market (GME, 2012). We assume that there are two different prices for each hour (peak and low price) (Table 7.2). The energy price is one of the inputs for the simulation, along with the probabilistic forecasts for PV generation.

TABLE 7.1

Technical and Economic Characteristics of the Power Plants

Power Plants	P_j^m (kW)	P_j^M (kW)	γ_{Pj} (c€)	β_{Pj} (c€)	α_{Pj} (c€)	β_{Bj} (c€)
Cogeneration (CHP)						
X_A: 60 kW	10	60	800	45.81	0.2222	
X_B: 60 kW	10	60	461	51.60	0.1000	
Y_A: 180 kW	36	180	892	34.40	0.0021	
Y_B: 180 kW	36	180	892	180	0.0420	
Traditional						
Z_A: 400 kW	80	400	1054	25.78	0.0005	
Z_B: 400 kW	80	400	1054	21.63	0.0025	
Boiler	0	4500				63.0

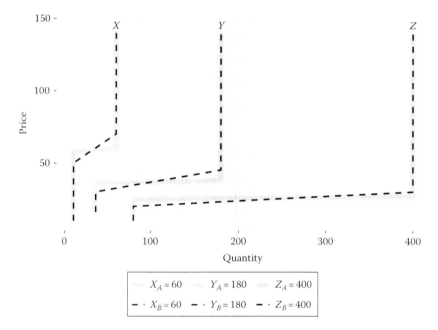

FIGURE 7.5
Power production related to various energy prices.

The forecast trends of power generation from the PV system are reported in Figure 7.6, where each panel shows the boxplots of the forecast for the power generated by the PV plant and computed by the AnEn algorithm.

The three curves resulting from the genetic algorithm optimization, shown in each panel, quantify the amount of PV electricity included in the

TABLE 7.2

Minimum and Maximum Spot Prices and Electrical Loads
Averaged over 12 Months for a 24-Hours Period

Hours	Min Price (€/MWh)	Max Price (€/MWh)	Load (kW)
1	30.7	102.6	440
2	25.7	96.6	440
3	21.4	92.0	440
4	17.3	87.0	440
5	14.9	85.7	440
6	16.6	86.8	740
7	16.1	85.5	1200
8	16.6	145.1	1905
9	26.4	188.8	2345
10	32.7	207.0	2405
11	32.2	207.1	2420
12	29.5	206.5	2440
13	27.2	143.9	2470
14	15.2	121.9	2465
15	12.1	144.5	2450
16	12.8	163.7	2395
17	20.2	186.6	2360
18	36.5	196.6	2335
19	56.9	222.3	1695
20	69.9	211.9	1425
21	64.1	324.2	1295
22	60.0	156.3	955
23	52.0	144.4	530
24	39.1	101.7	425

price-quantity bidding. These curves differ depending on the adversity to risk: solid (high risk), dotted (medium risk), and dashed (low risk). The three different risk-taking strategies are affected by the energy price.

In Figure 7.7, the electric power exchanged with the grid (green curve) and the electric power produced by traditional power plants and CHPs (blue sky curve) and PV system (violet curve), compared with the electric load profile, (red curve) are reported for different prices and risks. In fact, the manager of the grid can change the amount of energy that he or she needs and the power that he or she can produce, with the PV system too. The last is the function of the risk that he or she wants to sustain.

Figure 7.8 shows the optimal bidding curve at the 8th hour of the day for different risk-taking strategies of the prosumer.

The vertical and horizontal axes show the electricity prices and the electric power produced, respectively. The most difference between the three risk-taking strategies can be seen for low power values. The curve of the total

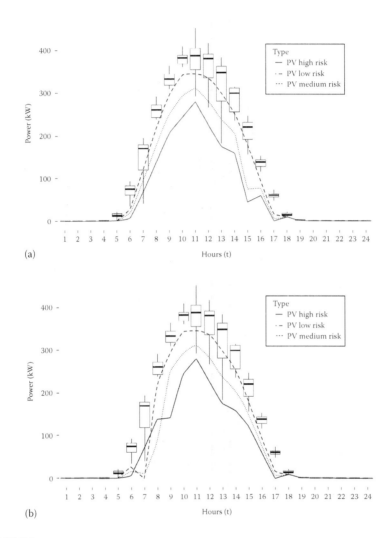

FIGURE 7.6
PV power production as a function of risk (adverse, neutral, incline) and price: (a) low price;
(b) medium price. The boxplots show the AnEn PV power forecasts, and the different curves
indicate the quantity of PV electricity included in the bidding depending on the prosumer
adversity to risk.   (*Continued*)

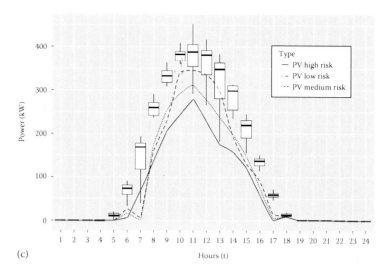

(c)

FIGURE 7.6 (*Continued*)
PV power production as a function of risk (adverse, neutral, incline) and price: (c) high price.
The boxplots show the AnEn PV power forecasts, and the different curves indicate the quantity
of PV electricity included in the bidding depending on the prosumer adversity to risk.

optimal production versus the spot price coincides with the curve of the
equivalent marginal cost of production. This curve is obtained by summing
for the same price the marginal costs of the various power generation units.
For each point of the bidding curve, the value of power offered is equal to
the difference between the total electric load requested and the total electric
power produced within the MG.

The hourly power offered in the day-ahead energy market coincides with
the power exchanged with the MV distribution network, in correspondence
to a specific energy market price. The power corresponding to the vertical
segment of the bidding curve is the difference between the load and the maximum production of the generating units compatible with the constraints,
including the energy produced by the PV system for the specific hour.

Results show that PV energy production can be integrated with optimal
outcomes in an MG if the prosumer strategy takes into account the uncertainty linked to the energy output. Outcomes show different optimal bids
depending on the risk adversity with respect to the uncertainty of PV power
production. The proposed methodology exhibits most improvement during
the hours in which the price of electricity is high and where the prosumer is
inclined to take risks.

FIGURE 7.7
Power production in function of risk in case of low price: (a) risk adverse, (b) risk neutral. Red line (top) is the electricity load, green line (medium) is the electricity power exchanged (buy/sell) with the MV grid, blue sky line (bottom) is the power generated. The last line is the total amount of energy produced by PV.   (*Continued*)

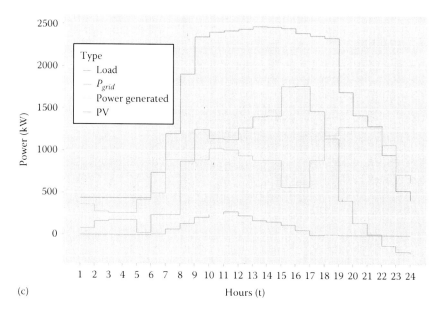

(c)

FIGURE 7.7 (Continued)
Power production in function of risk in case of low price: (c) risk incline. Red line (top) is the
electricity load, green line (medium) is the electricity power exchanged (buy/sell) with the MV
grid, blue sky line (bottom) is the power generated. The last line is the total amount of energy
produced by PV.

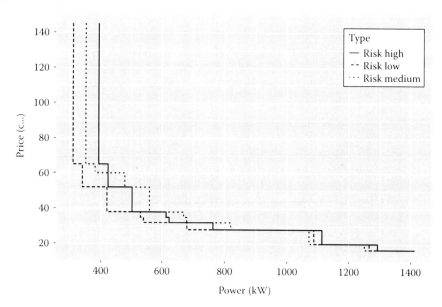

FIGURE 7.8
Bidding curve for the hour assuming a prosumer taking high risk (solid), medium risk (dotted),
and low risk (dashed).

7.6 Conclusion

A risk-bidding strategy for the day-ahead energy market was proposed to determine optimal economic choices for the management of a grid-connected microgrid comprising different generation units. It is assumed that the MG was controlled and managed by a prosumer that participates in the electricity market and needs to determine the optimal bidding.

Results show that PV energy production can be integrated with optimal outcomes in an MG if the prosumer strategy takes into account the uncertainty linked to the energy output.

PV power forecast and the uncertainty associated with its electricity generation have been evaluated through the AnEn approach.

Furthermore, outcomes show different optimal bids depending on the risk adversity with respect to the uncertainty of PV power production.

Further research will focus on the evaluation of the uncertainty of the energy price and also the electrical load, in order to provide to the manager of the MG a complete decisions support tool.

References

Alessandrini, S., F. Davo, S. Sperati, M. Benini, L. Delle Monache. 2014. Comparison of the economic impact of different wind power forecast systems for producers. *Advances in Science*, 11: 49–53.

Alessandrini, S., L. Delle Monach, S. Sperati, G. Cervone. 2015. An analogs ensemble for short term probabilistic solar power forecast. *Applied Energy*, 157: 95–110.

Alfredo, L., F. Jimenez, A.M. Jimenez, A. Falces, M. Mendoza-Villena, E. Garcia-Garrido, P.M. Lara-Santillan, E. Zorzano-Alba, P. J. Zorzano-Santamaria. 2012. Short-term power forecasting system for photovoltaic plants. *Renewable Energy*, 44(8): 311–317.

Bracale, A., P. Caramia, G. Carpinelli, A.R. Di Fazio, G. Ferruzzi. 2013. A Bayesian method for short-term probabilistic forecasting of photovoltaic generation in smart grid operation and control. *Energies*, 6: 733–747.

Bunn, D. 2000. Forecasting loads and prices in competitive power markets. *Proceedings of the IEEE*, 88(2): 163–169.

Cai, N., J. Mitra. 2010. A decentralized control architecture for a microgrid with power electronic interfaces. *Proceeding of the North American Power Symposium (NAPS)*, Arlington, TX, pp. 1–8.

Conejo, A.J., M. Carrion, J.M. Morales. 2010. Decision making under uncertainty in electricity market. In *International Series in Operation Research & Management Science*. Vol. 153, Springer, pp. 1–26.

Daoud, A.B., E. Sauquet, M. Lang, G. Bontron, C. Obled. 2012. Precipitation forecasting through an analog sorting technique: A comparative study. *Advances in Geoscience*, 29: 103–107.

Del Carpio, H.T.E., D.S. Ramos, R.L. Vasquez Arnez. 2010. Microgrid systems: Current status and challenges. *IEEE/PES Transmission and Distribution Conference and Exposition: Latin America (T&D-LA)*, Sao Paulo, 2010, pp. 7–12.

Delle Monache, L., F.A. Eckel, D.L. Rife, B. Nagarajan, K. Searight. 2013. Probabilistic weather prediction with an analog ensemble. *Monthly Weather Review*, 141: 3498–3516.

Dimeas, A., N. Hatziargyriou. 2004. A multi-agent system for microgrids. *Methods and Applications of Artificial Intelligence*, 3025: 447–455.

El-hawary, M.E. 2014. The smart grid. State of the art and future trends. *Electric Power Components and Systems*, 42(3): 239–250.

Favuzza, S. et al. 2006. Adaptive and dynamic ant colony search algorithm for optimal distribution systems reinforcement strategy. *Applied Intelligence*, 24: 31–42.

Feinberg, E.A. and D. Genethliou. 2005. Load forecasting. In *Applied Mathematics for Restructured Electric Power Systems: Optimization Control and Computational Intelligence*. Springer, Boston, MA, pp. 269–285.

Ferruzzi, G., G. Cervone, L. Delle Monache, G. Graditi, F. Jacobone. 2016. Optimal bidding in a day-ahead energy market for micro grid under uncertainty in renewable energy production. *Energy*, 106: 194–202.

Ferruzzi, G., F. Rossi, A. Russo. 2015. Determination of the prosumer's optimal bids. *International Journal of Emerging Electric Power Systems*, 16(6): 537–548.

Gestore del Mercato Energetico (GME). 2012. Testo integrato della Disciplina del Mercato Elettrico, Italy.

Graditi, G. et al. 2016a. Technical and economical assessment of distributed electrochemical storages for load shifting applications: An Italian case study, *Renewable and Sustainable Energy Reviews*, 57: 515–523.

Graditi, G. et al. 2016b. Energy yield estimation of thin-film photovoltaic plants by using physical approach and artificial neural networks. *Solar Energy*, 130: 232–243.

Hesham, K., M. Alfares, N. Mohammad. 2002. Electric load forecasting: Literature survey and classification of methods. *International Journal of Systems Science*, 33(1): 23–34.

Hu, L., G. Taylor, H.B. Wan, M. Irving. 2009. A review of short-term electricity price forecasting techniques in deregulated electricity markets. *Proceedings of the 44th International Universities Power Engineering Conference (UPEC)*, Glasgow, U.K., pp. 1–5.

Jain, A., A. Tuli, M. Kakkar. 2013. A review for electricity price forecasting techniques in electricity markets. *International Journal of Engineering Research & Technology*, 2(4): 714–1718.

Kantamneni, A., L.E. Brown, G. Parker, W.W. Weaver. 2015. Survey of multi-agent systems for microgrid control. *Engineering Applications of Artificial Intelligence*, 45(10): 192–203.

Kaur, A., J. Kaushal, P. Basak. 2016. A review on microgrid central controller. *Renewable and Sustainable Energy Reviews*, 55: 338–345.

Khan, A.R., A. Mahmood, A. Safdar, A.Z. Khan, N.A. Khan. 2006. Load forecasting, dynamic pricing and DSMin smart grid: A review. *Renewable and Sustainable Energy Reviews*, 54: 1311–1322.

Kumar, J.V. and D. Kumar. 2011. Optimal bidding strategy in an open electricity market using genetic algorithm. *International Journal of Advances in Soft Computing and its Applications*, 1(3): 1–14.

Kyriakides, E. and M. Polycarpou. 2007. Short term electric load forecasting: A tutorial. In K. Chen and L. Wang (eds.), *Trends in Neural Computation*. Springer, Berlin, Germany, pp. 391–418.

Lassater, B. 2001. Microgrid and distributed power generation. *IEEE Power Engineering Society Winter Meeting*. Conference Proceedings (Cat. No.01CH37194), Columbus, OH, 2001, pp. 146–149.

Luu, G.A. 2014. Control and management strategies for a microgrid. In *Electric Power*. Universite de Grenoble, Grenoble, France.

Mahmoud, S., S. Azher Hussain, M.A. Abido. 2014. Modeling and control of microgrid: An overview. *Journal of the Franklin Institute*, 351(5): 2822–2859.

Mongin, P. 1997. Expected utility theory. In Davis, J., Hands, W., and Maki, U. (eds.), *Handbook of Economic Methodology*, London, U.K., pp. 342–350.

Mureddu, M., G. Caldarelli, A. Chessa, A. Scala, A. Damiano. 2015 Green power grids: How energy from renewable sources affects networks and markets. *PLoS ONE*, 10(9): 1–15.

Ottensen, S.O., A. Tomasgard, S. Fleten. 2016. Prosumer bidding and scheduling in electricity markets. *Energy*, 94: 828–843.

Rossi, F. 2007. *Gestione dei sistemi elettrici nei mercati liberalizzati*, Vol. 41. ESI. Italy.

Schwaegerl, C., L. Tao, P. Mancarella, G. Strbac. 2009a. Can a microgrid provide a new paradigm for a network operation? An evaluation of their technical, commercial and environmental benefits. In *IET Conference Publications IET Conf Publ. 20th International Conference and Exhibition on Electricity Distribution*, CIRED 2009, Prague, July 1, 2009, pp. 1–4.

Schwaegerl, C., L. Tao, P. Mancarella, G. Strbac. 2010. A multi-objective optimization approach for assessment of technical, commercial and environmental performance of microgrids. *European Transactions on Electrical Power*, 2(1): 1269–1288.

Schwaegerl, C., L. Tao, J. Peças Lopes, A. Madureira, P. Mancarella, A. Anastasiadis, A. Krkoleva. 2009b. Can microgrids provide a new paradigm for network operation? An evaluation of their technical, commercial and environmental benefits. In *IET Conference Publications 20th International Conference and Exhibition on Electricity Distribution*, Prague, July 1, 2009, STREP project funded by the EC under 6FP, Report 668.

Shandurkova, I., A. Brendal, R. Bacher, S. Ottesen, A. Nilsen. 2012. A prosumer oriented energy market. NCE Smart Energy Markets.

Silva, M., H. Morais, Z.A. Vale. 2011. Distribution IMPROSUME. network short term scheduling in Smart Grid context. *IEEE Power and Energy Society General Meeting*, San Diego, CA, pp. 1–8.

Taheri, H., A. Rahimi-Kian, H. Ghasemi, B. Alizadeh. 2012. Optimal operation of a virtual power plant with risk management. *IEEE PES Innovative Smart Grid Technologies (ISGT)*, Washington, DC, pp. 1–7.

Timmerman, W., G. Huitema. 2009. Design of energy-management services supporting the role of the prosumer in the energy market. In *CEUR Workshop Proceedings*, Aachen, Germany.

Tsikalakis, A.G., N.D. Hatziargyriou. March 2008. Centralized control for optimizing microgrids operation. In *IEEE Transactions on Energy Conversion*, 23(1): 241–248.

Vogt, H., H. Weiss, P. Spiess, A.P. Karduck. 2010. Market-based prosumer participation in the smart grid. *Fourth IEEE International Conference on Digital Ecosystems and Technologies*, Dubai, United Arab Emirates, pp. 592–597.

Weron, R. 2014. Electricity price forecasting: A review of the state-of-the-art with a look into the future. *International Journal of Forecasting*, 30(4): 1030–1081.

8

Cost–Benefit Analysis for Energy Policies

Jacopo Torriti

CONTENTS

Abstract

Over the past three decades, cost–benefit analysis (CBA) has been applied to various areas of public policies and projects, including energy. Research on CBA varies significantly and can be classified into two wide areas of work: (1) studies identifying the technical and economic reasons underpinning CBA and (2) studies consisting of empirical evaluations over the performance of samples of CBA.

CBA is not the only example of economic tools applied to energy policy-making. Since the 1960s, the impact of energy policy measures has been assessed within the framework of various appraisal and evaluation tools. Decision analysis, environmental impact assessment, and strategic environmental assessment are all notable examples of appraisal tools predating and alternatives to CBA in the assessment of energy policies, programs, and projects. This chapter provides an overview not only of CBA but also of other appraisal and evaluation tools that have been historically applied to assess the impacts of energy policies, programs, and projects. It focuses on the types of data and models that typically inform CBAs for energy policies, the organizations involved, and issues of data exchange between energy companies and policy-makers. It is concluded that the technical and economic analyses underpinning CBAs on energy policy and regulation vary significantly depending on the type of policy, institutional aspects of decision-making, and availability of data.

Keywords: Cost–benefit analysis, energy economics, energy policy

8.1 Introducing Cost–Benefit Analysis

Over the past three decades, cost–benefit analysis (CBA) has been applied to various areas of public policies and projects, including energy. Research on CBA varies significantly and can be classified into two wide areas of work:

1. Studies that have attempted to define the technical–economic reasons underpinning CBA
2. Studies that carried out empirical evaluations over the performance of samples of CBA

From a theoretical point of view, CBA has been seen as a tool to increase the quality of regulation and public policy through welfare economics principles and Pareto efficiency. CBA in theory allows for the improvement of social and environmental conditions based on empirical evidence (Koopmans et al., 1964; Sunstein, 2002) while improving market competitiveness (Viscusi et al., 1987).

Empirical studies on CBA have focused on the choice of discount rate (Dasgupta, 2008; Gollier, 2002; Lind, 1995; Viscusi, 2007), the integration of distributional principles (Adler and Posner, 1999), the choice of data sets (Hahn and Litan, 2005; Morral, 1986), the performance of different methodologies for monetizing benefits, and costs in cases where a market value does not exist (Sunstein, 2004; Viscusi, 1988). The latter point is of particular interest given the distance between theory and practice and deserves further reflection.

The impact of energy policy measures has been assessed with various appraisal and evaluation tools since the 1960s. Decision analysis, environmental impact assessment (EIA), and strategic environmental assessment (SEA) are all notable examples of progenitors of CBA in the assessment of energy policies, programs, and projects. This chapter provides an overview not only of CBA but also of other policy tools that have been historically applied to assess the impacts of energy policies, programs, and projects. It focuses on the types of data and models that typically inform CBAs for energy policies, the organizations involved, and issues of data exchange between energy companies and policy-makers. Examples are derived from the European Commission, the United Kingdom, Italy, the Netherlands, and France.

Following the introduction, this chapter describes the historical development of CBA (Section 8.2) and classifies typologies of costs and benefits (Section 8.3). CBA is the most comprehensive of a family of economic evaluation techniques that seek to monetize the costs and/or benefits of proposals. Following standard classifications, benefits and costs can be broadly defined as anything that increases human well-being (benefits) or anything that decreases human well-being (costs). Section 8.4 defines efficiency under CBA. The chapter will then move the focus to the specific application of CBA for energy policies. It will do so by describing available policy tools to assess energy policies, programs, and projects (Section 8.5), identifying the institutions carrying out CBA on energy (Section 8.6), examining issues of data quality in energy policies (Section 8.7), before concluding (Section 8.8).

8.2 Historical Development of CBA

In order to understand the current scope and use of CBA in public policy-making (including in the energy sector), it is important to understand how this tool has developed over time. CBA was originally designed as an interface between engineering and economics in areas of civil engineering that relate to energy policies and projects. More precisely, Dupuit, a French engineer, and Marshall, a British economist, defined some of the formal concepts that are at the foundation of CBA. The Federal Navigation Act of 1936 required that the U.S. Corps of Engineers should carry out projects for the improvement of the waterway system when the total benefits of a project exceeded the costs. This was initiated by Congress, which ordered agencies to appraise costs and benefits when assessing projects designed for flood control as part of the New Deal.

In the 1950s, economists tried to provide a rigorous, consistent set of methods for measuring benefits and costs and deciding whether a project is worthwhile. This mainly consisted in applying compensation tests and distributional weights. However, such measures were considered by several economists as a failure (Adler and Posner, 1999). Notwithstanding opposition, in the United States, CBA was increasingly applied in an expanding domain of policy areas, often following the rationale that alternative policy appraisal tools were less efficient (Pearce and Nash, 1981).

Following some experiences in Scandinavian countries and Canada, the U.S. Executive Order 12291 of 1981 institutionalized CBA as a consistent method for the appraisal of government policies and regulations, hence marking the beginning of the CBA era (Posner, 2000). To date, there are soft-low requirements to conduct CBA on major policies and regulations in most OECD countries and examples from practice abound, as highlighted in the following.

8.3 Assessing Costs and Benefits

The theoretical and practical implications of assessing costs and benefits in CBA practice are similar for energy and nonenergy domains.

With regard to costs, each type of legislative change imposes various typologies of costs. Private companies, citizens, and public administration can be subject to an increase in costs. The first significant classification is with regard to private and societal costs. The former consist of what a citizen or household has to pay in relation to a legislative change. CBA is often used by public administrations as an instrument to measure only certain components of private costs. This is particularly the case when legislative change is expected to have impacts on individual categories of companies.

Social costs represent what society as a whole has to pay because of legislative change. They typically include negative externalities and exclude transfer costs among groups of citizens (or companies).

Figure 8.1 outlines the typologies of costs associated with legislative change. Costs for public administration mean management costs as well as enforcement costs, that is, costs associated with monitoring and inspections to ensure compliance. On the right of Figure 8.1, private costs are divided between costs for private citizens and private companies. The latter are broken down in terms of direct financial costs, administrative costs, capital costs, and efficiency costs.

With regard to benefits, the taxonomy presented in Figure 8.2 outlines some broad categories of benefits ordered from the highest level of monetization and quantification to the lowest.

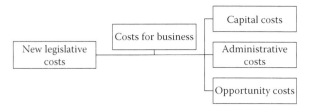

FIGURE 8.1
Typologies of costs in a public sector CBA.

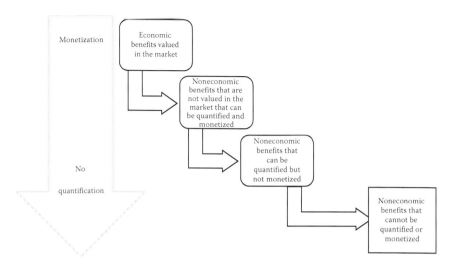

FIGURE 8.2
Classification of benefits in a public sector CBA.

Economic benefits for which a value is provided in the market are not difficult to monetize.

An example could come from a policy designed to add electricity generation from wind turbines. The market benefits are known because both the physical amount of energy that the extra turbines would provide (i.e., kWh) and the monetary value of the physical quantity (i.e., €/kWh) are known.

A contentious category of benefits consists of non-economic benefits that are not valued in the market, but can be quantified and monetized. There is no market value for this and neither for saving lives, but the monetary value of the benefit can be seen as a reduction in the risk of dying or catching a disease. Economists have developed four main methods for monetizing non-market values associated with reductions in risk:

1. Willingness to pay values ask citizens how much they would pay to reduce the likelihood of a specific risk. In practice, this is implemented through (a) stated preference surveys, where individuals are asked questions on changes in benefits; (b) close-ended surveys, where respondents are asked whether or not they would be willing to pay a particular amount for reducing risk; and (c) stochastic payment cards, which offer to respondents a list of prices and associates likelihood matrix describing how likely the respondent would agree to pay the various offered prices.

2. The human capital approach calculates the value of a human life saved, assessing the present value of the worker's earnings over the lifetime. The value is the benefit associated with reducing loss wages.

3. The cost of illness (or medical cost assessment) method consists of an estimate of the costs to the medical system for treatment due to illness.

4. Willingness to accept values are based on the wage premiums workers accept for risks. When the wage premium is divided by fatality risk, the result is the value of a statistical life saved.

Noneconomic benefits that can be quantified but not monetized include, for instance, the number of fish species saved from extinction. CBA is centered on human lives and impacts on other animal species are rarely taken into account in monetary terms as part of a policy appraisal, unless this refers specifically to ecological conservation and animal species protection. This is particularly the case for energy policies, which start from the assumption that energy is generated for and consumed by humans only. Nonetheless, handbooks on CBA including the British HM Treasury's (2012) *Green Book* may contain details about parameters to be used for plant species as part of the policy appraisal. The most recent version of the *Green Book* will contain a detailed discussion on the use of natural capital as part of CBA practice.

Benefits that cannot be quantified and, for the same reasons, cannot be monetized include predominantly several areas of social benefits. An example is the benefit of improving social justice, thanks to a new policy or regulatory change. There might be social indicators that address some of this change, but this is hardly reconciled to a monetary value. In the last U.S. Executive Order 13563 on CBA, it is stated that agencies should take account of "human dignity" and "fairness" values, although these are "difficult or impossible to quantify."

8.4 Basic Principles Underpinning CBA

Understanding how CBA works and its fundamental principles is vital in order to comprehend its application to energy policy. The core efficiency principles of CBA lie in welfare economics (i.e., the branch of economic theory that has investigated the nature of the policy recommendations that the economist is entitled to make) within the domain of allocative efficiency (Baumol et al., 1977; Perman, 2003). Allocative efficiency (i.e., allocation of scarce resources that gives maximum social well-being) is defined via the concept of a "Pareto improvement." A Pareto improvement is a reallocation of resources (e.g., a decision to develop) that makes at least one individual better-off without making anyone worse-off. Pareto efficiency/optimality is achieved when it is impossible to make one individual better-off without worsening the condition of at least one other individual.

If economists restricted their domain of advice to Pareto improvements, they would not be able to advise on much as most decisions involve a trade-off between making someone better-off at the expense of making someone else worse-off. What could be required is that the individual who gains must compensate the individual who loses, for all of the latter's loss. If the individual who gains still gains after having paid out the compensation in whatever way (e.g., cash), the move would still be a Pareto improvement. This is not much less restrictive, because actual compensation is rarely paid. As an alternative, economists developed the idea of potential Pareto improvements. The Kaldor compensation test (after Nicholas Kaldor) sanctions a move from one allocation of resources to another, if the winner could compensate the loser and still be better-off. In this case, the compensation does not actually have to be paid. Hicks identified a problem with Kaldor's compensation test—namely, that it could sanction a move from one allocation to another, but it could equally sanction a move in the opposite direction, depending on where the problem starts. Instead, Hicks suggested that the loser could compensate the winner for forgoing the move, without being worse-off than if the change took place.

A later paper by Scitovsky (1951) unraveled the problem. Both rules need to be satisfied, such that a reallocation is desirable if, on the one hand, the winners could compensate the losers and still be better-off and, on the other hand, the losers could not compensate the winners for the reallocation not occurring and still be as well-off as they would have been if it did occur.

Because CBA is based on the Kaldor–Hicks efficiency criterion, it means that the benefits should be enough that those that benefit could in theory compensate those that lose out. It is justifiable for society as a whole to make some worse-off if this means a greater gain for others. Under Pareto efficiency, an outcome is more efficient if at least one person is made better-off and no one is made worse-off. This is a stringent way to determine whether or not an outcome improves economic efficiency. However, some believe that in practice, it is almost impossible to take any social action, such as a change in economic policy, without making at least one person worse-off (Buchanan, 1959). Using Kaldor–Hicks efficiency, an outcome is more efficient if those that are made better-off could in theory compensate those that are made worse-off, so that a Pareto-improving outcome results. An allocation is defined as "Pareto efficient" or "Pareto optimal" when no further Pareto improvements can be made.

In the case where the public sector supplies goods, CBA becomes a tool for judging efficiency (Stiglitz, 2000). The concept of efficiency, though, is normally thought on the premise of the market economy. This is particularly controversial when the decision being contemplated involves some cost or benefit, for which there is no market value or which, because of an externality, is not fully reflected in the market value.

8.5 Beyond CBA: Other Appraisal Tools to Assess Energy Policies, Programs, and Projects

CBA is a widespread tool in the domain of energy policy appraisal techniques, as discussed earlier. However, there are other ways to feed evidence into the formulation of energy policies, programs, and projects. This section provides a historical overview of alternative economic appraisal tools for energy policies, programs, and projects.

In the 1960s, decision analysis was first applied to investigate problems in oil and gas exploration. Its application was consequently extended from private to the public sector (Huang et al., 1995). An early 1990s review enumerates 86 decision analysis studies that appeared in peer-reviewed journals from 1970 to 1989 (Corner and Kirkwood, 1991). Subsequent surveys found that decision analysis was frequently used to address strategic or policy decisions related to energy, such as investment options facing the utility industry, choice between different energy technologies, synthetic fuel policy, commercialization of solar photovoltaic systems, management of nuclear waste, and acid rain control (Zhou et al., 2006). Decision analysis focuses on resolving conflicts between objectives, dealing with uncertainty about the outcomes, and appraisal of multiple options. Given the fact that energy and environmental issues are generally complex and inevitably involve multiple objectives, the techniques involved in decision analysis varied depending on the level of uncertainty associated with the specific policy or project, with multicriteria analysis cited more frequently.

Environmental issues became increasingly central to the development of energy policies and to the activities of the energy industry in the 1970s. This led to the upsurge of EIA and (later) SEA. The former was first a commonly accepted practice when developing energy infrastructure and then became a regulatory requirement in several legal dominions around the world (Petts, 1999). The latter became an established practice in the late 1990s and over time improved its legal status in some jurisdictions (Dalal-Clayton and Sadler, 2005).

Environmental impacts of individual project proposals are the main focus of EIAs. Examples consist of new power plants or new hydroelectric dams (Mirumachi and Torriti, 2012). Legislation was first introduced in the United States in 1969. In Europe, EIAs were introduced, thanks to the EC Directive in 1985 (85/337), which was amended in 1997 (97/11). Currently, EIAs are carried out by institutions like the World Bank, the OECD member states, transition countries, and several developing countries.

Because the environmental performance of the energy sector has been subject to higher degrees of scrutiny, questions were raised about whether EIA was the right tool to address the challenges associated with energy supply (Wood, 2003). SEA is designed to address environmental issues at a higher level of planning, which may take place at a regional, national, and supernational scale. This is consistent with the idea that environmental protection needs to be embedded into energy frameworks at early phases of conception. The main origins of SEA in the EU relate to the Strategic Environmental Assessment Directive (2001/42). SEA is supposed to complement EIA for strategic actions. Strategic action is a more nebulous process than the formal submission of a development proposal, as in EIAs. Thus, SEAs address concepts rather than particular activities and must deal with incremental and nonlinear policy processes (Wood and Dejeddour, 1992). Because it is focused on strategic actions, SEA is designed to include a stronger consideration of alternative options than EIA. The environment is often singled out in SEA, more so than in EIA or CBA, in large part because of the need to bolster its importance relative to the economic and social dimensions (Thérivel and Partidário, 1996). Finnveden et al. (2003) note that it is not clear which, if any, applications within the energy sectors require an SEA, and Jay (2010) notes that SEA has not been extensively adopted in the area of energy production.

This may be explained in relation to the fragmented nature of the industry, since generation, transmission, distribution, and supply operate as separate markets—at least where liberalization took place. This makes the use of strategic planning tools more difficult. Today, SEA has potential in the fields of landscape, carbon reduction, and air quality.

8.6 CBAs on Energy: Institutional Differences

8.6.1 CBAs on Energy Policies by Government Departments

Because of the importance of energy for fuelling economic growth, energy policy sits firmly in any governmental agenda. The institutionalization of energy policy often translates into the presence of energy ministries or energy departments within government (Newbery, 1989). These, like any other government department, are charged with the task of formulating their policy with the support of CBAs. In addition, energy-related policies can be developed within departments for environment, industry, and transport. To date, no research has collected and let alone examined the body of CBAs produced by government departments. This section seeks to capture four salient features of CBAs by government departments. Four issues characterize the CBA in energy policies by government departments:

1. Unlike CBAs by regulatory authorities, which feature a high level of techno-economic analysis (see following section), governmental CBAs on energy policy tend to follow a more generalist approach. CBAs conducted by government departments are often less quantitative in terms of the analysis and geared to a less specialist audience. An example of this is the CBA on the Green Deal and Energy Company Obligation in the United Kingdom (DECC, 2011). The Green Deal aims to overcome access to capital and mismatched incentive problems. The Energy Company Obligation aims to provide additional support to deliver socially cost-effective measures that are not likely to be taken up under current policies and provides measures to relieve fuel poverty. In essence, these are complementary policy measures intended to address barriers to the slow uptake of cost-effective energy efficiency measures. The government's CBA estimates that the Green Deal will lead to 125,000–250,000 households being lifted out of fuel poverty by 2023, but there have been criticisms with the way this figure was derived. More specifically, it has been argued that the CBA is too simplistic. For example, it neglects distributional issues: the Green Deal might increase fuel poverty since the policy might only benefit better-off end users and not be supportive of the fuel poor (Arie, 2012). The CBA was also criticized for applying very high discount rates (7%, which is significantly higher than other policies). Indeed, the CBA shows that investments do not generate positive net present values for discount rates of 5%. However, other similar case studies show that the types of technological solutions contained in the Green Deal would create negative net present values even with discount rates as low as 1.5% (Energy Saving Trust, 2010).

2. There is a tendency to outsource research and analysis for those CBAs that require highly specialized electrical engineering and energy economics knowledge. For instance, in Ireland, Spain, and the United Kingdom, consultants such as Frontier Economics, London Economics, Mott MacDonald, NERA, and RedPoint have been contracted to carry out CBAs and to come out with policy options for CBAs on key policy areas such as smart metering, energy efficiency, renewable heat incentives, feed in tariffs, and so on. The smart meter CBA by Mott MacDonald highlighted that the most advanced smart metering options would have negative net present values. It was noted that this initial negative assessment was partly

duc to assumptions that limit the value of demand side management. Hence, the final CBA on smart meters presents a preferred rollout option with a positive net present value (DECC, 2009).

3. The tendency to delegate pieces of analysis also results in interest groups gathering in specialist groups to produce the quantitative sections of a CBA. For instance, as part of the UK Government Electricity Market Reform, DECC asked a technical experts group, comprising the UK transmission system operator (National Grid), distribution network operators, and energy aggregators, to produce analysis regarding the details of transitional arrangements to include demand side response and energy storage within newly formed capacity mechanisms. Similarly, in the case of the UK policy for "zero carbon homes," which is part of a more general approach to low-energy buildings, leadership on CBA was given by the UK Government to the Sustainable Building Task Group in 2003 (BIS, 2008). The group was cochaired by the Chairman of the Environment Agency and the Deputy Chairman of English Partnerships and consisted of representatives from industry and environmental groups. According to Hauf (2012), a similar combination of government, industry, and environmental groups united in the same policy formulation body also occurred in France. The specific proposals for the amendment of building regulations were developed by the Building Regulations Advisory Committee, which produced the results of the accompanying CBA.

4. The high political implications of energy policies mean that there are occasions where policies are pushed forward regardless of the "better regulation" principles dictated by the same government. For instance, in 2011, the initiative by DECC of rolling out smart meters was the only example of policy escaping the one-in, one-out rule applied by the UK coalition government. According to the one-in, one-out rule, regulation whose direct incremental economic cost to business and civil society organizations exceeds its direct incremental economic benefit to business and civil society organizations can only come to place along with deregulatory measures whose direct incremental economic benefit to business and civil society organizations exceeds its direct incremental economic cost to business and civil society organizations (HM Treasury, 2011). In other words, the rule requires that no new national regulation is brought in without other regulation being cut by a greater amount. This also implies that the introduction of new regulations and removal of existing regulations are both government interventions and require their own separate CBA. The smart meters' initiative was classified as an "in" under the one-in, one-out methodology, because the CBA showed some £57 million equivalent annual net cost to business. However,

it was introduced as new legislation without any significant "outs." DECC stated that there was plan to simplify the nuclear decommissioning financing and fees framework—hence reducing paperwork burdens on operators (DECC, 2012).

8.6.2 CBAs on Energy Regulation

Since the 1990s in several developed and developing countries, the liberalization of energy markets has been coupled with the emergence of energy regulatory agencies. CBA, along with stakeholder consultation, has become a common tool in liberalizing or liberalized markets. In some countries, the energy regulators stand out as a positive exception for having implemented CBA more rigorously than other government departments and other agencies (Renda, 2004). A review of CBA implementation across Italy confirms that the gas and electricity regulatory authority follows appropriate criteria (La Spina and Cavatorto, 2008).

Regulatory authorities have been driven by the dual aim of (1) reducing prices to end users and (2) improving the quality of energy supply. In order to obtain lower prices for end users, one of the main tasks of energy regulators is to regulate the prices of distribution companies, because these are considered regional monopolies and need incentives to ensure that they improve efficiency and raise the quality of supply. Energy regulators use price control reviews to regulate the prices that distribution companies can charge suppliers for transporting electricity through their networks (Cowell, 2004). The reviews normally take place every 4–5 years and involve a complex methodology that delivers data supporting the CBA.

The approach for the CBA commences with companies submitting a business plan setting out their operating costs, proposals for improving quality of supply, and capital expenditure estimates for the next 5 years. The regulators typically enter these data into a series of cost benchmarking exercises, with companies' estimated expenditure benchmarked against each other. Given the importance of price control reviews in determining the development of distribution electricity systems, the accompanying CBA is arguably the most significant piece of regulatory analysis in any liberalized energy market. For this reason, two issues are particularly worth noting:

1. In spite of its highly technical features, the final CBA seldom contains much detail about the actual cost curves of distribution network operators. Issues of competition mean that regulators may not make explicit allowances for particular infrastructural projects. Thus, some companies may enter dialogue with the regulators during the review process but find relatively limited justifications for the review decisions in the final CBA (Guy and Marvin, 1996).

2. Price review CBAs typically neglect non-techno-economic impacts, including social and environmental impacts. For instance,

Ofgem's (2001) Environmental Action Plan affirms that "the choices made in the design of price control regulation can have wide ranging environmental impacts" but specifies what has long been the regulator's position on environmental and social matters: that it is not an environmental policy-maker and does not produce social policy.

Regulatory authorities cyclically conduct another example of major CBA on Quality of Supply regulation, which has similar rules in various European countries (e.g., United Kingdom, France, and Italy). In Italy, this is subject to 4-yearly revisions as part of which the regulatory authority sets the penalties and incentives for distribution network operators. The CBA process has been studied as an example of effective integration of various factors, including economic analysis based on end users' willingness to pay for better energy provision (Ajodhia et al., 2006) and consultation (Fumagalli and Lo Schiavo, 2009). Torriti et al. (2009) describe the CBA process for two reviews of the Quality of Supply regulation in terms of preliminary analyses, research studies, alternative regulatory options, consultation, and CBA. They highlight some of the analytical and procedural issues typically associated with CBA: the creation of alternative options, the development of cost–benefit analysis, the disparity between analytical effort and available resources, and the need to communicate in an informed manner with interested stakeholders. The experience from this example also shows that when attention is paid to these details, CBA can generate unexpected results.

8.6.3 CBAs by the European Commission

A significant share of the European Commission's policies is in the area of energy, several of which require an impact assessment and, consequently, a CBA. Figure 8.3 illustrates the number of CBAs conducted by different European Commission Directorate Generals (DGs) on energy policies between 2003 and 2013. The CBAs on energy in Figure 8.3 were sourced from the official website of the European Commission (www.europa.org).

Before the creation of DG Energy in 2010, its predecessor—DG for Transport and Energy (TREN)—conducted most of the CBAs on energy. However, over the years, CBAs on energy policies have been carried out also by DG Environment, Climate, Informatics (DIGIT) and Economic and Financial Affairs (ECFIN). After the year 2009, the intensification of policy-making activity around climate change targets, with renewable energy, energy efficiency, and lighting policies, justifies the higher number of CBAs.

The level of analysis varied over time. What Hanley and Spash (1993) stated at the beginning of the 1990s, that in the area of energy and environment benefits have not always been well integrated into the European Community policy assessments, cannot apply to present times. Individual CBAs received some attention by researchers with regard to their economic

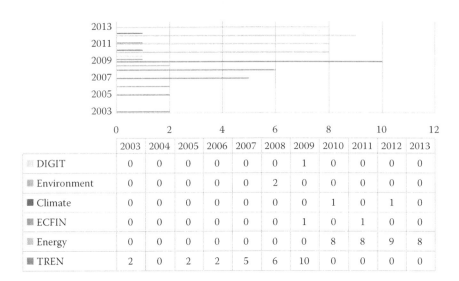

	2003	2004	2005	2006	2007	2008	2009	2010	2011	2012	2013
DIGIT	0	0	0	0	0	0	1	0	0	0	0
Environment	0	0	0	0	0	2	0	0	0	0	0
Climate	0	0	0	0	0	0	0	1	0	1	0
ECFIN	0	0	0	0	0	0	1	0	1	0	0
Energy	0	0	0	0	0	0	0	8	8	9	8
TREN	2	0	2	2	5	6	10	0	0	0	0

FIGURE 8.3

Cost–benefit analyses on energy policies carried out by the different Directorates General of the European Commission (2003–2013).

assessments. For instance, the CBA on the EU's Objectives on Climate Change and Renewable Energy for 2020 (European Commission, 2008) considers the actual EU partitioning between Emission Trading Scheme (ETS) and non-ETS sectors to be cost-effective, whereas, according to Böhringer et al. (2009), the CBA did not take into account the excess costs associated with differential emission pricing (Torriti, 2010).

8.7 Quality of the Data and CBAs on Energy

One of the main issues around CBAs on energy regards the extent to which data from energy companies are used as part of the appraisals. Most of the energy regulators monitor the markets in order to foresee critical issues, prevent disturbances, and enact timely regulatory actions. To ensure this task, they need significant amounts of data from market actors. Ideally, CBAs on energy could contain all the available information on market performance, compare operations over time and across markets, publicly release all data submitted to and produced by the market and system operators, and even anticipate instances where small market flaws may develop into market failures. However, there are at least three obstacles to the transparent exchange of information between energy companies and energy regulators.

First, there are data that companies have a right to maintain confidential. Second, there are data that are public in principle but costly to gather and assemble. Third, there is no general consensus on the desirability of data exchange. Ofgem's guidance on CBA admits the challenges of capturing competition effects of regulatory change.

In the case of markets being opened up to competition, for example, it is inherently difficult to predict with any accuracy the potential efficiency benefits that introducing a competitive process might bring, or to quantify meaningfully the dynamic benefits of competition such as the scope for increased innovation and the introduction of new products, services and technologies. (Ofgem, 2013, p. 23)

In principle, the exchange of information from energy companies to policy-makers is desirable because lack of exchange of data leads to incomplete information and inefficient screening of the market (Brown, 2001). According to this view, under publicly available CBAs, the information these disclose should be available not only for those who have the legal and financial capability to access data but also to all market and nonmarket actors. However, in practice, incumbents argue that there are instances where transparency may violate property rights or harm business when the disclosure of crucial information can alter the competitive process (Campbell and Lindberg, 1990). An example of this relates to the treatment of electricity consumption data from smart meters. In principle, if regulators had access to such data, they could make informed decisions about tariffs, based on actual end users' consumption. In reality, the consultation and CBAs conducted in the Netherlands show that both energy companies and consumers were opposed to the disclosure of consumption data (Cavoukian et al., 2010). Moreover, even new entrants to the market may find data exchange problematic. Perfect visibility of the strategies of competitors may be beneficial to the defense of market power by the dominant company. Some delay in making market bids transparent may help the strategies of new competitors. In the recent Electricity Market Reform in the United Kingdom, DECC proposed to publish historical data on the bidding prices for the Short Term Operating Reserve, a service for the provision of additional active power from generation and/or demand reduction, and energy aggregators, which have only entered energy markets in the past 5 years, opposed such change.

Compared with the European Commission's DG Energy, the U.S. Federal Energy Regulatory Commission features a higher legislative power for access to data and a greater financial capacity to purchase data. Since data are kept confidential by the regulator, business concerns of being negatively affected by data disclosure for competition purposes are limited. This arrangement implies that in the United States the transparency of CBAs is sacrificed in support of effective market monitoring and higher quality of data and analysis.

In the EU, institutional market monitoring activities are lagging. In some European countries, the regulator is recipient of a wealth of data (Gilbert et al., 2002), and the main issue is whether to publish them in a CBA or not.

In other countries, data gathering does not represent a problem, but organizational deficiencies make the treatment of data rather difficult.

8.8 Conclusion

This chapter described the context, theories, and main features of and alternatives to CBA applied to energy policies. The discussion on the historical development of CBA shows that this appraisal tool represents a mix of engineering and economics methodologies that are very suitable to most energy policy problems. CBA is most directly applicable to civil engineering projects and programs that rely on quantifiable units of cost. The chapter provides a breakdown of typologies of costs and benefits, hence noting some of the difficulties associated with the practice associated with benefits appraisal in energy problems. Most economic appraisal techniques tend to treat net benefits (or undiscounted cash flows) as given. However, gathering data on future flows of benefits is an intrinsically uncertain exercise. CBAs on energy policies have traditionally relied on different modes of data collection and statistical inference, based among other things on the type of institution conducting the economic appraisal (e.g., central government departments, European Commission, or regulatory authorities).

This is one of the reasons why the chapter addresses what economic efficiency means for CBAs on energy policies. This chapter reviewed economic efficiency as the core principle underpinning CBA, as it is a method by which this concept of efficiency can be applied to publicly supplied goods. Assuming that energy policies and projects bring utility to people, these people are associated with a willingness to pay, which represents the benefit of supplying such goods. Willingness to pay corresponds to a willingness to accept (i.e., the cost for supplying the goods). In energy policy appraisals, difficulty exists, however, in transferring the efficiency concept for market goods to publicly supplied goods. Two relevant points involve the efficiency concept in CBA: (1) energy policies may be concerned with a much broader range of consequences than firms and (2) energy suppliers may not always use market prices in evaluating projects either because the market prices may not exist or because market prices may not represent true marginal social benefits/costs. The efficiency criteria on which CBAs for energy policies are based are relatively different from the traditional willingness to pay *vis-à-vis* willingness to accept relation because they are supposed to take into account distribution and equity issues. When efficiency conflicts with other values, it is actually impossible to create economic welfare criteria that integrate all values. An example of this discussion on the impractical use of CBA in the context of the macroeconomics of energy policy comes from projects aimed to address fuel poverty. These create both winners and losers. In most of the cases, losers already belong to the poorest and more marginalized members of society. While fuel poverty projects can bring enormous benefits to society, their costs to the poorest have effects on their health and even their lives (Kanbur, 2002). The use of CBA by government departments for fuel poverty projects is widespread and often criticized for not considering areas like basic needs approaches, shadow prices, social discount rates, and macroeconomic shocks to public goods (Brent, 1998; Devarajan et al., 1997; Kirkpatrick and Weiss, 1996).

The examples given in this chapter offer a picture of the range of CBAs applied to energy policies. This chapter has observed how the level of techno-economic analysis varies in CBAs conducted by regulators, government departments, and the European Commission. However, no judgment is placed here on the value of less technical CBAs. Indeed, a lower level of analysis may yield positive benefits in terms of greater engagement with stakeholders and the wider public in the development of policies that will have significant consequences for society as a whole. This benefit is even greater when taking into account the detachment of the lay public from energy policies (Cotton and Devine-Wright, 2012). Unlike some of the more technically focused exercises that have been used to assess energy regulation, CBA for energy policy is intended to be an inclusive and participative process, in which there is an opportunity for deliberation and consensus building. It has been pointed out elsewhere that the application of cost–benefit analysis to energy policies emphasizes three typical weaknesses: (1) the exclusive concern with economic values, (2) the treatment of uncertainty, (3) and the neglect of intergenerational effects (Simpson and Walker, 1987).

Very much like environmental policies, energy policies tend to be designed to achieve multiple objectives ranging from climate change to utilities' tariffs. Correspondingly, the impacts generated by energy policies tend to vary substantially in nature and size. Over the years, policy-makers have expanded the types of analytical tools used in the appraisal and evaluation of energy policies from narrowly scoped geophysical and ecological models, on the one hand, and purely socioeconomic-oriented tools of decision analysis, on the other hand, to highly integrated assessment tools, such as CBA. Through an expansion of geographic and temporal scopes and depiction of large complex systems, CBA models were expected to overcome many of the shortcomings of earlier analyses, that is, absent or inadequate depiction of technological change, micro-behaviors of economic actors (e.g., firms and consumers), intergenerational trade-offs and fairness, and uncertainty (Greening and Bernow, 2004). However, given their broad approach, CBAs are prone to the problem of finding a balance between quantification of current economic and physical phenomena and future variations in the supply and demand of energy systems. In energy policy, the main added value of the CBA system has been associated with explicitly providing a range of policy-relevant criteria including a broad range of stakeholder opinions that can be used to assess (in traditional cost–benefit terms) and develop alternative environmental and energy policies.

References

Adler, M. and Posner, E. (1999). Rethinking cost-benefit analysis. *Yale Law Journal*, 109, 165–247.

Ajodhia, V., Lo Schiavo, L., and Malaman, R. (2006). Quality regulation of electricity distribution in Italy: An evaluation study. *Energy Policy*, 34, 1478–1486.

Arie, S. (2012). Understanding the risks of the Green Deal. Working Paper, Smith School of Enterprise and the Environment, University of Oxford. http://www.smithschool.ox.ac.uk/people/SamArie_WorkingPaper.pdf, accessed June 1, 2016.

Baumol, W. J., Bailey, E. E., and Willig, R. D. (1977). Weak invisible hand theorems on the sustainability of multiproduct natural monopoly. *The American Economic Review*, 67, 350–365.

BIS (2008). Sustainable Buildings Task Group. http://webarchive.nationalarchives.gov.uk/+/http://www.berr.gov.uk/sectors/construction/sustainability/sbtg/page11919.html, accessed June 1, 2016.

Böhringer, C., Löschel, A., Moslener, U., and Rutherford, T. (2009). EU climate policy up to 2020: An economic impact assessment. *Energy Economics*, 31, S295–S305.

Brent, R. J. (1998). *Cost-Benefit Analysis for Developing Countries*. Edward Elgar Publishing Ltd, Cheltenham.

Brown, M. A. (2001). Market failures and barriers as a basis for clean energy policies. *Energy Policy*, 29(14), 1197–1207.

Buchanan, J. M. (1959). Positive economics, welfare economics, and political economy. *Journal of Law and Economics*, 2, 124.

Campbell, J. L. and Lindberg, L. N. (1990). Property rights and the organization of economic activity by the state. *American Sociological Review*, 55, 634–647.

Cavoukian, A., Polonetsky, J., and Wolf, C. (2010). Smart privacy for the smart grid: Embedding privacy into the design of electricity conservation. *Identity in the Information Society*, 3(2), 275–294.

Corner, J. L. and Kirkwood, C. W. (1991). Decision analysis applications in the operations research literature, 1970–1989. *Operations Research*, 39(2), 206–219.

Cotton, M. and Devine-Wright, P. (2012). Making electricity networks "visible": Industry actor representations of "publics" and public engagement in infrastructure planning. *Public Understanding of Science*, 21(1), 17–35.

Cowell, R. (2004). Market regulation and planning action: Burying the impact of electricity networks?. *Planning Theory and Practice*, 5(3), 307–325.

Dalal-Clayton, B. and Sadler, B. (2005). *Strategic Environmental Assessment: A Source-Book and Reference Guide to International Experience*. Earthscan, London, U.K.

Dasgupta, P. (2008). Discounting climate change. *Journal of Risk and Uncertainty*, 37, 141–169.

DECC (2009). Impact assessment of a GB-wide smart meter roll out for the domestic sector. www.decc.gov.uk/assets/decc/consultations/smart%20metering%20 for%20electricity%20and%20gas/120090508152831e@@smartmeteCBAdomestic.pdf, accessed June 1, 2016.

DECC (2011). The Green Deal and Energy Company Obligation Impact Assessment. https://www.gov.uk/government/uploads/system/uploads/attachment_ data/file/43000/3603-green-deal-eco-ia.pdf, accessed June 1, 2016.

DECC (2012). DECC's 4th Statement of New Regulation. https://www.gov.uk/ government/uploads/system/uploads/attachment_data/file/48461/5787-decc-4th-statement-of-new-regulation.pdf, accessed June 1, 2016.

Devarajan, S., Squire, L., and Suthiwart-Narueput, S. (1997). Beyond rate of return: reorienting project appraisal. *The World Bank Research Observer*, 12(1), 35–46.

European Commission (2008). Impact assessment, document accompanying the package of implementation measures for the EU's objectives on climate change and renewable energy for 2020, SEC(2008) 85/3, January 23, 2008. European Commission, Brussels, Belgium.

Finnveden, G., Nilsson, M., Johansson, J., Persson, Å., Moberg, Å., and Carlsson, T. (2003). Strategic environmental assessment methodologies— Applications within the energy sector. *Environmental Impact Assessment Review*, 23(1), 91–123.

Fumagalli, E. and Lo Schiavo, L. (2009). Regulating and improving the quality of electricity supply: The case of Italy. *European Review of Energy Markets*, 3(2), 1–27.

Gilbert, R., Neuhoff, K., and Newbery, D. (2002). Mediating market power in electricity networks. Working Paper, Department of Economics, University of Berkeley, Berkeley, CA.

Gollier, C. (2002). Discounting an uncertain future. *Journal of Public Economics*, 85, 149–166.

Greening, L. A.and Bernow, S. (2004). Design of coordinated energy and environmental policies: Use of multi-criteCBA decision-making. *Energy Policy*, 32(6), 721–735.

Guy, S. and Marvin, S. (1996). Disconnected policy: The shaping of local energy management. *Environment and Planning C*, 14, 145–158.

Hahn, R. and Litan, R. (2005). Counting regulatory benefits and costs: Lessons for the US and Europe. *Journal of International Economic Law*, 8(2), 473–508.

Hanley, N. and Spash, C. L. (1993). *Cost-Benefit Analysis and the Environment*. Edward Elgar, Cheltenham, U.K.

Hauf, H. (2012). A collaborative approach to policy: A case study on the Zero Carbon Hub: 2020 Public Services Hub at the RSA. http://www.zerocarbonhub. org/sites/default/files/resources/reports/Zero_Carbon_Hub_Progress_ Report_29th_May_2012.pdf, accessed June 1, 2016.

HM Treasury (2011). One-In, One-Out (OIOO) methodology. HM Treasury. https:// www.gov.uk/government/uploads/system/uploads/attachment_data/ file/31616/11-671-one-in-one-out-methodology.pdf, accessed June 1, 2016.

HM Treasury (2012). Accounting for environmental impacts: Supplementary Green Book guidance. HM Treasury, London, U.K.

Huang, J. P., Poh, K. L., and Ang, B. W. (1995). Decision analysis in energy and environmental modeling. *Energy*, 20(9), 843–855.

Jay, S. (2010). Strategic environmental assessment for energy production. *Energy Policy*, 38, 3489–3497.

Kanbur, R. (2002). Economics, social science and development. *World Development*, 30(3), 477–486.

Kirkpatrick, C. H. and Weiss, J. (Eds.) (1996). *Cost-Benefit Analysis and Project Appraisal in Developing Countries*. Edward Elgar Publishing, Cheltenham, U.K.

Koopmans, T. C., Diamond, P. A., and Williamson, R. E. (1964). Stationary utility and time perspective. *Econometrica*, 32, 82–100.

La Spina, A. and Cavatorto, S. (2008). *Le autorità indipendenti*, Vol. 573. Il Mulino, Bologna, Italy.

Lind, R. (1995). Intergenerational equity, discounting, and the role of cost-benefit analysis in evaluating global climate policy. *Energy Policy*, 23, 379–389.

Mirumachi, N. and Torriti, J. (2012). The use of public participation and economic appraisal for public involvement in large-scale hydropower projects: Case study of the Nam Theun 2 Hydropower Project. *Energy Policy*, 47, 125–132.

Morral, J. (1986). A review of the record. *Regulation*, 2, 25–34.

Newbery, D. (1989). Energy policy issues after privatization, in *The Market for Energy*. D. Helm, J. Kay, and D. Thompson (Eds.). Clarendon Press, Oxford, U.K., pp. 30–54.

Ofgem (August 2001). *Environmental Action Plan*. Ofgem, London, U.K.

Ofgem (2013). Guidance on Ofgem's Approach to Conducting Impact Assessments. Proposed Guidance. https://www.ofgem.gov.uk/ofgem-publications/82590/proposedguidanceonofgemsapproachtoconductingimpactassessments.pdf, accessed June 1, 2016.

Pearce, D. W. and Nash, C. A. (1981). *The Social Appraisal of Projects: A Text in Cost-Benefit Analysis*. Macmillan, London, U.K.

Perman, R. (Ed.) (2003). *Natural Resource and Environmental Economics*. Pearson Education, London, U.K.

Petts, J. (Ed.) (1999). *Handbook of Environmental Impact Assessment*. Blackwell, Oxford, U.K.

Posner, R. A. (2000). Cost-benefit analysis: Definition, justification, and comment on conference papers. *The Journal of Legal Studies*, 29(S2), 1153–1177.

Renda, A. (2004). Qualcosa di nuovo nell'AIR? Riflessioni a margine del dibattito internazionale sulla better regulation. *L' industCBA*, 26, 331–364.

Scitovsky, T. (1951). The state of welfare economics. *The American Economic Review*, 41, 303–315.

Simpson, D. and Walker, J. (1987). Extending cost-benefit analysis for energy investment choices. *Energy Policy*, 15(3), 217–227.

Stiglitz, J. E. (2000). Capital market liberalization, economic growth, and instability. *World Development*, 28(6), 1075–1086.

Sunstein, C. (2004). *The Cost-Benefit State: The Future of Regulatory Protection*. American Bar Association, Chicago, IL.

Thérivel, R. and Partidário, M. R. (1996). *The Practice of Strategic Environmental Assessment*. Earthscan Publications Ltd., London, U.K.

Torriti, J. (2010). Impact assessment and the liberalisation of the EU energy markets: Evidence based policy-making or policy based evidence-making?. *Journal of Common Market Studies*, 48(4), 1065–1081.

Torriti, J., Fumagalli, E., and Lo Schiavo, L. (2009). L'AIR nella pratica di una Autorità indipendente: l'esperienza dell'Autorità per l'energia elettrica e il gas di applicazione dell'AIR alla regolazione della qualità del servizio. *Mercato Concorrenza Regole*, 10(2), 285–321.

Viscusi, W. K. (1988). Irreversible environmental investments with uncertain benefit levels. *Journal of Environmental Economics and Management*, 15, 147–157.

Viscusi, W. K. (2007). Rational discounting for regulatory analysis. *University of Chicago Law Review*, 74, 209–246.

Viscusi, K., Magat, W., and Huber, J. (1987). An investigation of the rationality of consumer valuations of multiple health risks. *Rand Journal of Economics*, 18, 465.

Wood, C. (2003). *Environmental Impact Assessment: A Comparative Review*. Pearson Education.

Wood, C. and Dejeddour, M. (1992). Strategic environmental assessment: EA of policies, plans and programmes. *Impact Assessment*, 10(1), 3–22.

Zhou, P., Ang, B. W., and Poh, K. L. (2006). Decision analysis in energy and environmental modeling: An update. *Energy*, 31(14), 2604–2622.

9

Benchmarking Energy Efficiency Transitions in MENA Countries

Tareq Emtairah and Nurzat Myrsalieva

CONTENTS

9.1 Introduction: Background and Context

Pursuing energy efficiency as a national policy objective is a cross-cutting challenge for governments. The outcomes expressed in proxies such as *energy intensity* of the economy and/or *energy productivity* are linked to so many variables and interventions operating at different levels; from macroeconomic conditions, technology development and path dependency, energy supply and pricing regimes to influencing the behavior of millions of energy consumers (see for e.g., Howarth and Andersson 1993; Rosenberg 1994, p. 161; Biggart and Lutzenhister 2007; Gillingham et al. 2009). Therefore, a wide range of policies and measures have been used by governments to facilitate the transition toward improved energy efficiency.

Given the possible range of governmental interventions and the special context of each country, an effective transition regime for energy efficiency might appear difficult to characterize at first glance. However, in a global review covering close to 110 countries conducted by the International Energy Agency (IEA) in 2010, the authors conclude that *energy efficiency policy is more likely to be successful if an effective system of energy efficiency governance is established* (OECD/IEA 2010, p. 14). This implies that it is not only the policies that matter but also the governance arrangements for enabling energy efficiency transitions. In the same IEA report, a grouping of the possible arrangements is made in three main headings: *enabling frameworks, institutional arrangements, and coordinating mechanisms* (OECD/IEA 2010, p. 15).

Informed by this understanding of energy efficiency transition and its governance arrangements, and while working for the Regional Center for Renewable Energy and Energy Efficiency (RCREEE),[*] the authors participated in the development and testing[†] of a benchmarking index for a systematic analysis of energy efficiency transitions in RCREEE's Arab member states within the Middle East and North Africa (MENA) region. The underlying logic and motivation for RCREEE is that an energy efficiency governance perspective is warranted for the adequate assessment of countries' transitional processes. An assessment of this nature would also allow the Center to gauge the extent to which political rhetoric and targets are realistic within the national market conditions, institutional arrangements, and national capacities (Myrsalieva and Samborsky 2013; Myrsalieva and Barghouth 2015). Furthermore, it should facilitate strategic conversations with national focal points working with energy efficiency questions on the strengths and weaknesses of national energy efficiency strategies.

This chapter presents the experience and lessons learned from this benchmarking exercise. It is organized to give a brief overview of the drivers for energy efficiency in the MENA countries, followed by a description of the benchmarking framework and its parameters, hereafter referred to as the Arab Future Energy Index (AFEX). Finally, key findings and insights based on the process and outcomes are presented and discussed.

9.2 Drivers of Energy Efficiency in MENA

With a few exceptions, the majority of Arab countries in MENA have been late in paying adequate attention to energy efficiency (Emtairah and Chaaban 2013). In the past, governmental efforts in most countries of the region can be characterized at best as ad hoc, uncoordinated, and mostly driven by donor-countries' projects.

---

[*] RCREEE was founded in 2008 through a cooperation agreement among 10 Arab countries with the support from the German and Danish governments. A Secretariat was set up in Cairo in 2010 and given the mandate by its founding members to act as policy advocacy partner to national governments and support the member states with their efforts in developing markets for renewable energy and energy efficiency investments.

[†] The authors also acknowledge the contribution and involvement of many other actors in the actual production of the benchmarking index from within and outside RCREEE.

Not more than 5 years ago, a noticeable shift in attitude occurred among national authorities and key stakeholders within the energy sectors in favor of coordinated efforts toward energy efficiency. This shift is driven by the convergence of several internal and external factors putting pressure on national governments to adequately balance the supply and demand in energy services.

The most important factors often cited in this discussion include

1. High rates of population growth and lifestyle changes leading to above-average demand growth for energy services particularly in the electricity sector
2. Steep fluctuations in primary energy prices coupled with a wide range of universal subsidies leading to pressures on national budgets to adequately finance capacity expansion, while at the same time sustain artificially low prices for energy services

The effect of these key factors and the resulting dynamics pressuring the internal energy systems and their policy implications are not necessarily the same due to the fact that the energy and economy profiles of the MENA countries differ considerably (ESMAP 2009). Further treatment of these differences and the drivers is provided in a later section of this chapter; however, it needs to be emphasized that from a fundamental cause–effect relationship, these two factors remain significant driving forces in the discourse on energy efficiency transitions across all countries in the region. Other factors can be cited such as aging and inefficient infrastructure and technologies across the energy system value chain but tend to be more specific to one country or a smaller group of countries. In comparison to other regions, climate change and environmental consideration played rather in an significant role* in the discourse and subsequent strategies shaping the energy efficiency transitions in the region (Reiche 2010; Waterbury 2013).

With this background, noticeable efforts from countries in the region toward improved energy efficiency are taking shape. The formal adoption in 2011 by the Arab Ministerial Council for Electricity of the recommendations and supporting guidelines encouraging member countries to establish *national energy efficiency action plans* (NEAPS) and provided further impetus toward more coordinated and systematic efforts to organize energy efficiency strategies and programs at the national level. Still, the landscape remains unclear as to how far the efforts in the member countries in terms of policies, strategies, and implementation are adequately guiding the transition toward the right direction.

---

* Despite courteous reference to these considerations in public policy documents.

9.3 The Arab Future Energy Index

To understand better energy system challenges and guide the energy transition process in the MENA region, RCREEE in 2013 launched the Arab Future Energy Index (AFEX) as a policy assessment and benchmarking tool. Adding to what has been stated earlier on the underlying aim from the benchmarking; in the communication with the member states, AFEX is promoted as a tool to provide *consistent and fact-based analysis of energy transition processes in the MENA region through benchmarking exercise considering the political and economic realities of the countries* (Myrsalieva and Samborsky 2013).

The assessment is based on the compilation and analysis of detailed, country-specific data according to the set of predefined indicators and parameters. The idea would be to publish AFEX on a regular basis to carefully monitor the changes and the progress in the region in the field of renewable energy and energy efficiency. The benchmarking models for energy efficiency and for renewable energy, while following the same structure, differ in the assessment goal and types of parameters.* In this chapter, when references are made to AFEX, they refer to the energy efficiency component.

9.4 Areas of Assessment and Parameters

The energy efficiency component of AFEX provides a comparative overview of the current state of energy efficiency in the MENA region and presents an analysis of countries' performance across four evaluation criteria: (1) energy pricing, (2) policy framework, (3) institutional capacity, and (4) utility.

The energy pricing evaluation criteria assess the structure of electricity tariffs, gasoline, and diesel, including the energy subsidy reform efforts. The policy framework evaluation criteria assess the countries' level of commitment to overcome market, social, and political barriers to energy efficiency by formulating and adopting strategies, policies, and target-based action plans. The institutional capacity evaluation criterion assesses the capacity of governmental stakeholders to design, implement, and evaluate energy efficiency policies. Under utility criteria, the efficiency of the power supply is assessed including power generation efficiency and efficiency of transmission and distribution networks. Table 9.1 shows how the four criteria are broken into 10 factors, which are further broken into sets of quantitative and qualitative indicators.

The evaluation criteria, factors, and indicators have been chosen based on the assessment of the main drivers, barriers, and applicable international

---

* The assessment reports are available through www.rcreee.org.

TABLE 9.1

AFEX Energy Efficiency Indicators

Category	Factor	Indicator	Score/Measuring Unit
Energy pricing	Electricity price structure	Time-of-use price structure	Number of segments
		Other price-based incentives	Number of incentives
	Energy subsidies	Electricity subsidies in residential sector	Subsidy percentage (benchmarked to Palestinian retail prices for electricity)
		Electricity subsidies in commercial sector	Subsidy percentage (benchmarked to Palestinian retail prices for electricity)
		Electricity subsidies in industrial sector	Subsidy percentage (benchmarked to Palestinian retail prices for electricity)
		Subsidies for gasoline	Retail price per liter
		Subsidies for diesel	Retail price per liter
Policy framework	Energy planning	Energy strategy with long-term EE objectives	Officially adopted; nonexistent
		National Energy Efficiency Action Plan	Adopted; not adopted; under development
		Quantitative, time-bound EE targets for residential, tertiary, industrial, transport, and utility sectors	Adopted; not adopted; under development
	Regulatory framework	Framework legislation for EE measures	Adopted; draft prepared; nonexistent
		EE regulations for buildings	Mandatory; voluntary or under preparation; nonexistent
		Statutory obligation to install solar water heaters in new buildings	Adopted; under preparation; nonexistent
		Mandatory minimum energy performance standards with labeling schemes for household appliances	Number of appliances
		Regulatory phase-out of inefficient lighting technology	Policy adopted; under preparation; nonexistent
		EE regulations for industries	Number of regulatory policies
		Regulatory phase-out of old or inefficient vehicles	Adopted; under development; nonexistent
		Policies discouraging car ownership and promoting public transport and car sharing	Number of policies adopted

(Continued)

TABLE 9.1 (Continued)

AFEX Energy Efficiency Indicators

Category	Factor	Indicator	Score/Measuring Unit
	Financial incentives	EE Fund	Established by law; sources of financing are clear; disbursement procedure is clear; fund is operational
		Internal tax benefits	Number of tax incentives
		Customs duty for CFL and LED bulbs	Customs rate percentage
		Customs duty for solar water heaters	Customs rate percentage
Institutional capacity	EE Institutions	Designated EE units within ministries	Presence of designated EE unit; adequacy of technical and human resources; capacity to formulate and monitor EE policies
		Designated EE agency	Presence of designated EE agency; adequacy of technical and human resources; capacity to implement EE policies
	Implementation capacity	Number of EE buildings built	Percentage of new building stock built according to EE regulations
		Number of demonstration projects	Expert assessment from 0 to 10 based on number of demonstration projects; market size of construction industry
		Solar water heater diffusion rate	m^2 of panels per 1000 inhabitants
		Corruption Perception Index	CPI score
Utility	Generation	Power generation efficiency	National average percentage
		Share of renewable energy in generation mix	Percentage of total installed capacity and electricity generated
	Transmission and distribution	Transmission and distribution losses (including technical and commercial losses)	National average percentage

best practices to improving energy efficiency performance in the region. One of the major barriers identified at the outset of the project included heavily subsidized energy prices, which were prevalent in both energy-exporting and energy-importing countries. Other barriers included lack of energy efficiency policies, lack of targets, weak institutions to lead energy efficiency policies, lack of awareness, and lack of capacities and expertise. The drivers for energy efficiency differed between energy-importing and energy-exporting countries. For energy-importing countries, energy security, reducing budget deficits, and reducing reliance on imported energy products were the main considerations for pursuing energy efficiency. For energy-exporting countries, energy efficiency was viewed as a way to preserve diminishing natural resources, to comply with international commitments under climate change policy, and to show leadership and goodwill.

The main constraint for selecting the indicators was data availability. Lack of detailed data availability on energy consumption constitutes one of the major challenges in most MENA countries. Many countries in the region lack a centralized office responsible for collecting detailed energy-related data. Often data are scattered between different institutions, inconsistent, or simply unavailable. Taking this situation into account, the indicators had to be practical, operational, relevant to the region, resource-efficient, and easy to measure.

9.5 Comparing Results from Two Cycles of Benchmarking

9.5.1 2013 Cycle

Under AFEX Energy Efficiency 2013, 13 MENA countries were assessed and ranked.* The MENA region comprises politically, economically, and socially diverse countries. Although the 13 countries represent only part of the whole MENA region, these countries, nevertheless, collectively cover geographical area of 8,886,000 km^2 with a total population of 301 million people. The largest country by population is Egypt with about 83 million people and the smallest country is Bahrain with only 1.3 million people. The region includes countries with versatile attributes related to energies and economy status: some are energy-dependent (Morocco, Tunisia, Jordan), and others are energy-independent (Algeria, Bahrain, Libya, Iraq) some are categorized as upper high income (Bahrain), and some lower middle income (Yemen, Sudan).

---

* Countries assessed under AFEX Energy Efficiency 2013: Algeria, Bahrain, Egypt, Iraq, Jordan, Lebanon, Libya, Morocco, Palestine, Sudan, Syria, Tunisia, and Yemen.

Just as the countries within the region are different, their performance in energy efficiency is also different. The most diverse performance has been observed under the energy pricing category. Although the MENA region as a whole is characterized by heavily subsidized energy prices, the differences between individual countries' energy prices are significant. Figure 9.1 shows that the highest energy prices are in energy-dependent countries: Palestine, Morocco, Tunisia, and Jordan. Residential electricity tariffs in Palestine are nearly 20 times higher than residential tariffs in Bahrain, Syria, and Iraq. Industrial electricity tariffs in Morocco shown in Figure 9.2 are five times higher than industrial tariffs in Libya, Egypt, Iraq, and Bahrain. Subsequently, the countries with less subsidized energy prices ranked the highest under energy pricing evaluation criteria.

With regard to energy efficiency policies, the countries again showed differences. Five out of thirteen countries (Egypt, Lebanon, Palestine, Sudan, and Tunisia) have national energy efficiency action plans in place with specific energy savings targets. Other five countries (Algeria, Jordan, Morocco, Syria, and Tunisia) have energy efficiency legislation in place. Only one country (Tunisia) has officially banned the import and sale of incandescent light bulbs and only Tunisia has the most comprehensive program in place to promote industrial energy efficiency. Only six countries have put in place mandatory energy efficiency regulations for buildings. The customs duties on the import of efficient light bulbs and solar water heaters ranged from 0% in Jordan to 30% in Algeria.

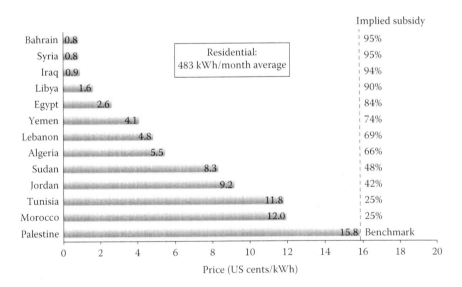

FIGURE 9.1
Residential electricity tariffs and implied subsidies (2011).

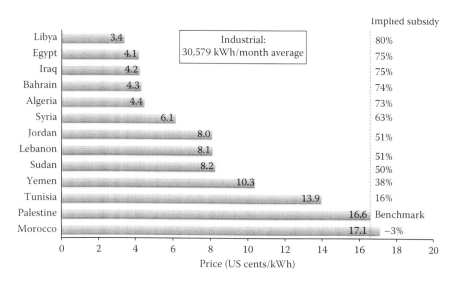

FIGURE 9.2
Industrial electricity tariffs and implied subsidies (2011).

The overall performance of the region under policy framework evaluation criteria was found weak. Many countries still had to improve their regulatory frameworks. Often, countries have incomplete policies (building code covering only thermal envelope in Bahrain or framework energy efficiency law without implementing decrees in Morocco), policies covering only certain sectors (energy savings targets for the residential sector only in Egypt), conflicting policies (free distribution of efficient light bulbs while maintaining high import duty on the same efficient light bulbs in Algeria), no energy efficiency policies at all (Libya, Yemen), or policies comprising of only voluntary schemes (Lebanon and Iraq). Only one country, Tunisia, stood out with a comprehensive regulatory framework covering all aspects of the economy and including a wide range of policy measures (mandatory, voluntary, and incentive-based).

Under institutional capacity category, the countries showed the poorest performance. Again, only Tunisia stood out with a relatively better implementation rate of energy efficiency policies. This is mostly due to the presence of a strong dedicated national energy efficiency agency—ANME. In most countries, energy efficiency policies greatly suffered from lack of enforcement. This was mostly due to the fact that countries lacked strong institutional base to implement energy efficiency policies.

This finding reinforces the message from the OECD/IEA report (2010) that the success of energy efficiency policies is conditional upon effective energy efficiency governance structure. The heart of an energy efficiency governance system lies in a strong dedicated energy efficiency agency with the capability to design, formulate, implement, and evaluate energy efficiency

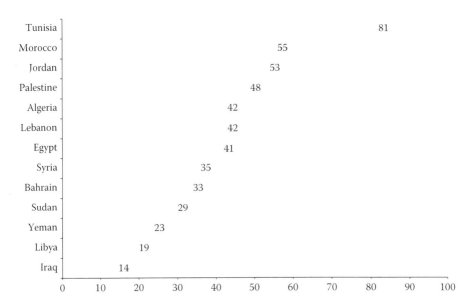

FIGURE 9.3
Country ranking based on 2013 assessment.

policies and programs. Such an agency also needs to be capable of coordinating activities among various stakeholders to ensure more effective use of human, capital, and technical resources in achieving energy efficiency objectives. Most countries in MENA region lack this effective energy efficiency governance system, which in our assessment drives the overall poor performance in energy efficiency. Often countries either have understaffed energy efficiency agency (Egypt), or agency without any policy formulation powers (Jordan), or uncoordinated efforts of multiple agencies (Bahrain, Sudan), or no dedicated energy efficiency agency at all (Iraq).

Figure 9.3 shows the final ranking of countries under the AFEX Energy Efficiency 2013 assessment. In the final ranking, Tunisia emerges as the leader, followed by Morocco, Jordan, and Palestine.

9.5.2 2015 Cycle

The scope of AFEX Energy Efficiency 2015 was broadened. Four more countries, Kuwait, Qatar, Saudi Arabia, and UAE, have been added to the assessment. Also, the scope of assessment have been broadened to include the assessment of energy efficiency in the transport sector. Although the overall ranking of countries did not change much, many developments took place under individual indicators since the publication of AFEX 2013.

The biggest developments took place under the energy pricing category. Six countries, Egypt, Jordan, Morocco, Tunisia, Sudan, and Yemen, implemented energy subsidy reform efforts. Egypt introduced a 5-year plan on gradual increases of electricity prices and significantly increased prices of gasoline, diesel, and natural gas. Jordan also approved a 5-year plan for gradual increases of electricity tariffs and completely removed subsidies from oil products. Morocco also eliminated subsidies for gasoline and industrial fuel. Tunisia introduced increases on electricity tariffs and fuel prices. Sudan and Yemen both have significantly increased prices for gasoline and diesel.

The policy framework category also showed some developments. Three more countries adopted national energy efficiency action plans with specific energy savings targets (Iraq, Jordan, and Tunisia adopted its third plan). The biggest improvement has been made in phasing out inefficient appliances. Three countries, Jordan, UAE, and Qatar, introduced technical standards for home appliances with minimum energy efficiency requirements. Bahrain, UAE, and Qatar also introduced a ban on the import and sale of inefficient light bulbs. The two large energy-consuming sectors were found to be the least regulated across all MENA countries: transport and industrial sectors. These two sectors represent areas with great untapped energy efficiency potential. The transport sector specifically appears to be the most challenging to tackle as the energy efficiency improvements are associated with high investment costs for infrastructure development projects.

The least progress has been observed under the institutional capacity category. The countries overall did not make much progress in strengthening the institutional base and improving the implementation capacity. The greatest challenge with enforcement lies in the building sector. Although the countries put in place energy efficiency building regulations, many of these regulations remain unenforced. To improve the compliance in this sector, countries need to dedicate more effort to establishing clear and transparent enforcement procedures, building technical capacities, and designing measures to promote voluntary compliance.

Figure 9.4 illustrates the performance of countries over two categories: policy framework and institutional capacity. In this figure, only Tunisia and UAE appear to perform well under both categories. These countries have better policy frameworks and relatively stronger institutional capacities, enabling better implementation of energy efficiency policies and measures. Jordan scores well under policy framework, but needs to improve its institutional capacity to ensure effective implementation of its policies. Majority of the countries are still in the lower left quadrant. This means that countries still need to improve their regulatory frameworks and strengthen their institutional capacities.

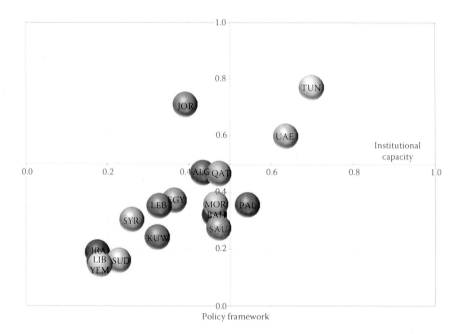

FIGURE 9.4
Countries' performance over two categories based on 2015 assessment.

9.6 Discussion and Concluding Remarks

Initially, RCREEE developed the benchmarking tool as a means to understand better energy transition processes in the MENA region and as a means to guide its intervention strategies. The tool successfully served this purpose and became an avenue for sharing regional experiences and for engaging in broader discussions and partnerships. One example is the case of Tunisia. Through AFEX it became evident that Tunisia has achieved great results in improving energy efficiency in the country and its national energy efficiency agency played a key role in this process. This triggered interests from several other countries to learn about the Tunisian experience and institutional arrangements. As a result, RCREEE organized several study tours and exchange missions to Tunisia.

AFEX has also become an important information platform, which served not only policymakers, but also development organizations and other stakeholders. RCREEE was able to collect good set of data points and consolidate across specific indicators. Even when there were no data points, proxies served well the purpose. Example is the approximate subsidies solution as a practical and sufficient proxy to characterize realities on the ground. Estimating exact amount of subsidies is a challenging task due to a big range of subsidies, different modes of implementation, poor data quality, limited data availability, and lack of transparency. Given the lack of data on subsidy rate in all Arab countries and difficulties in measuring subsidies according to one common methodology, a proxy was developed to estimate energy subsidy levels. The results were cross-checked with individual national experts and seemed to have correlated with the reality. This gave RCREEE experts sufficient grounds to have strategic conversations with the relevant authorities and stakeholders about the difficult subject of subsidies and build necessary arguments for change.

Like any other benchmarking tool, AFEX created a competitive attitude among the countries. This was particularly evident with higher-ranking countries. AFEX also became somewhat a common regional avenue to display countries' efforts in promoting sustainable energy developments in their countries. For instance, a representative from one country argued that the ranking does not reflect the level of efforts the agency in that country is putting toward promoting energy efficiency; however, on closer examination between the assessment parameters and the activities of the country, it was possible to have a constructive dialogue comparing stand-alone ad hoc promotional programs to measures affecting market transformation.

The authors of the benchmarking tool expected a considerable challenge on the selection and type of parameters and indicators, but surprisingly, very little of that came out from the counterparts in the member countries. Most of the discussions were mainly about the results of the findings, about general trends, about progress of individual countries, and very little about the actual choice of indicators.

Overall, the benchmarking provided a basis to have more structured discussions about specific challenges and problems. For example, there are two indicators for assessing energy efficiency in the building sector: (1) whether the countries have put in place energy efficiency regulations for buildings; and (2) the percentage of new building stock built according to the energy efficiency regulations. The results of these two indicators showed that many countries have put in place energy efficiency building regulations, but the enforcement was completely lacking. These findings elicited more focused discussions on the specific challenges with enforcement mechanisms.

A side effect is that having a consistent and comprehensive regional scale monitoring of the energy efficiency transitions can be of valuable guidance to donor-driven programs in the recipient countries. The experience from AFEX suggests that such an index can lower the transaction costs of designing effective donor support programs at national or regional scales.

Given the heterogeneity of national context, the exercise demonstrated again that countries can take different paths to improving energy efficiency. If energy prices are heavily subsidized, there is still ample space to advance energy efficiency investments through, for example, putting in place stringent mandatory energy efficiency regulations and taking a lead in the government sector. Consistent also with empirical observations elsewhere, the institutional arrangements for coordination and enforcement are absolutely instrumental to improving energy efficiency.

_____

References

Biggart, L. and N. Lutzenhister. 2007. Economic sociology and the social problem of energy inefficiency. *American Behavioural Scientist* 50(8): 1070–1087.

Emtairah, T. and F. Chaaban. 2013. Energy efficiency. In: Abdel Gelil, I., M. El-Ashry, and N. Saab, eds. *Sustainable Energy: Prospects, Challenges, and Opportunities.* Beirut, Lebanon: Arab Forum for Environment and Development. Available online: http://www.afedonline.org/report2013/english/5-eng.pdf. Accessed June 15, 2016.

ESMAP. 2009. Tapping a hidden resource energy efficiency in the Middle East and North Africa. Washington, DC: Energy Sector Management Assistance Program: World Bank Group. Available online: http://siteresources.worldbank.org/INTMNAREGTOPENERGY/Resources/MENA_Energy_Efficiency_2009.pdf. Accessed June 15, 2016.

Gillingham, K., R. Newell, and K. Palmer. 2009. Energy efficiency economics and policy: Annual review of resource economics. *Annual Reviews* 1(1): 597–620.

Howarth, R. and B. Andersson. 1993. Market barriers to energy efficiency. *Energy Economics* 15(4): 262–272.

Myrsalieva, N. and A. Barghouth. 2015. Arab future energy index: Energy efficiency. Cairo, Egypt: The Regional Center for Renewable Energy and Energy Efficiency. Available online: http://www.rcreee.org/content/arab-future-energy-index%E2%84%A2-afex-2015-energy-efficiency. Accessed June 15, 2016.

Myrsalieva, N. and B. Samborsky. 2013. Arab future energy index: Energy efficiency. Cairo, Egypt: The Regional Center for Renewable Energy and Energy Efficiency. Available online: http://www.rcreee.org/sites/default/files/rcreee_reportsstudies_afex_ee_report_2012_en.pdf. Accessed June 15, 2016.

OECD/IEA. 2010. *Energy Efficiency Governance.* Paris, France: IEA. Available online: http://www.iea.org/publications/freepublications/publication/eeg.pdf. Accessed June 07, 2016.

Reiche, D. 2010. Energy policies of Gulf Cooperation Council (GCC) countries—Possibilities and limitations of ecological modernization in rentier states.*Energy Policy* 38(5): 2395–2403.

Rosenberg, N. 1994. *Exploring the Black Box: Technology, Economics, and History.* Cambridge, U.K.: Cambridge University Press.

Waterbury, R. 2013. The political economy of climate change in the Arab region. Arab Human Development Report Research Paper Series 2. Regional Bureau for Arab States: United Nations Development Programme. Available online: http://www.arab-hdr.org/publications/other/ahdrps/AHDR%20ENG%20Waterbury%20v3.pdf. Accessed June 18, 2016.

10

Analysis of the European Energy Context: A Snapshot of the Natural Gas Sector

Vincenzo Bianco

CONTENTS

10.1 Introduction

Since its discovery in the United States at Fredonia, New York, in 1821, natural gas has been used as fuel in areas immediately surrounding gas fields. In the period 1920–1930, some long-distance pipelines were installed in the United States to transport gas from remote areas to industrial centers (Ikoku, 1984).

The natural gas industry of today did not emerge until after World War II, when the consumption started to increase rapidly. This growth was due to several factors, including the development of new markets, replacement of coal as a fuel for providing space and industrial process heat, strong demand for low-sulfur fuels that emerged in the mid-1960s, and so on (Ikoku, 1984).

In light of this, natural gas can be considered a relatively new fuel, even though it plays an important role in the global energy balance, and its impact in the European energy scenario is significant.

In particular, from the end of the 1990s, a sharp increase of consumption has been detected due to its massive utilization for power generation in combined cycle gas turbine (CCGT) plants.

The use of natural gas for power generation has been encouraged by the implementation of the EU Emission Trading Scheme (i.e., a scheme for the trading of CO_2 emission allowances), which gives advantages to low-carbon-intensive fuels.

Due to the efficiency and coverage of the distribution network, natural gas is largely used also in other sectors of activity, for example industrial or residential, for different purposes, such as process steam generation, heating, and so on.

Natural gas is considered as the necessary fuel to support the European energy transition toward renewable energies; therefore, it has a deep strategic relevance in the EU energy policy. However, the indigenous production is limited and able to satisfy only a partial share of the demand.

To support the consumption of natural gas, different infrastructures were developed and others are under development, in order to connect the EU with exporter countries (i.e., Russia, Algeria, and others) by means of pipelines. These infrastructures rigidly connect supply and demand; therefore, to limit the market power of the suppliers, many regasification plants have been also built all around Europe to permit the import of liquefied natural gas (LNG) from other regions (i.e., Nigeria, Venezuela, and others), allowing more competitive supply sources.

In the last years, due to the economic downturn, a decrease of natural gas consumption has been detected, causing a destabilization of the market, and it is not clear if it will be a transient effect or if it will lead to a reshape of the market.

This chapter will analyze all the main issues connected with natural gas sector in Europe, by taking into account the current situation and possible future perspectives.

10.2 Preliminary Background

Natural gas is a combustible mixture of hydrocarbon gases. In general, it is mainly formed of methane (70%–90%), but it often includes ethane, propane, butane, and pentane (0%–20%), and a minor share of impurities such as carbon dioxide, nitrogen, and hydrogen sulfide.

The composition of natural gas can widely vary according to the location of extraction.

It can be divided in three main categories, namely, dry natural gas, wet natural gas, and sour natural gas:

1. Dry natural gas is a purified product represented almost entirely by methane.
2. Wet natural gas is natural gas that contains methane and other hydrocarbons.
3. Sour natural gas is natural gas that contains a large amount of hydrogen sulfide. This is an issue because of the corrosive effects and the formation of SO_x during combustion.

Natural gas is a fossil fuel. This means that it is the result of the decomposition of plants, animals, and microorganisms that lived millions of years ago. It represents organic material prevented from its complete decay.

Natural gas originates from two different mechanisms, namely, thermogenic and biogenic.

1. Thermogenic methane: it is formed from organic particles covered by sediment, mud, and debris that form an increasingly thicker layer on top of the organic matter. This sediment exerts a high pressure on the organic matter, which compresses it. The joint combination of pressure and temperature (temperature increases as one gets deeper in the soil) allows to break the carbon bonds in the organic matter, which leads to the formation of methane.
2. Biogenic methane: it is formed as a consequence of the action of methanogens—microorganisms able to produce methane. Methanogens chemically break down organic material, leading to the formation of methane. They can be found in areas close to the earth surface where oxygen is not present; therefore, the methane produced in this way is found close to the earth surface and, often, it is lost in the atmosphere. Sometimes, the methane remains confined in the underground and it can be extracted.

Natural gas can be classified in two macro-categories, namely, conventional and unconventional. As given in Wang et al. (2014), a natural gas deposit can be defined as "conventional" if it is contained in rocks (often limestone or sandstone) with a permeability of more than 1000 microdarcy that have interconnected spaces that allow the gas to flow freely in the rock and to well boreholes. On the other hand, natural gas is classified as unconventional gas if it is situated in rock formation with a permeability of less than 1 millidarcy, which makes the gas difficult to flow.

The definition of "unconventional" gas based on a single value of permeability is of limited significance. In fact, the commercial completions of conventional basins may be achieved when the permeability is in the microdarcy range.

Another definition of unconventional can be given from the economic point of view. According to this, unconventional gas is natural gas that cannot be produced at economic flow rates or in economic volumes of natural gas, unless the well is stimulated by a large hydraulic fracture treatment (Wang et al., 2014).

There are four types of unconventional gas: shale gas, coal-bed methane, tight gas, and gas hydrates.

1. Shale gas: the gas is in shale deposits typically found in river deltas, lake deposits, or flood plains. Shale is both the source and the reservoir for the natural gas. This can either be "free gas," which is trapped in the pores and fissures of the shale rocks, or adsorbed gas, which is contained in surfaces of the rocks (Wang et al., 2014).
2. Coal-bed methane: in coal deposits, significant amounts of methane-rich gas are generated and stored within the coal structure when it has an extremely low permeability.
3. Tight gas: unlike shale gas or coal-bed methane, it is formed outside the rock formations where it has migrated over millions of years into extremely impermeable hard rock or sandstone or limestone formations that are unusually nonporous (Wang et al., 2014).
4. Gas hydrates: natural gas hydrates (also known as clathrates) are solid gas molecules surrounded by a lattice of water molecules. They are formed by water and natural gas (methane) at high pressures and low temperatures (ETSAP, 2010).

As it can be imagined, the extraction of unconventional gas is more complex than the conventional one, due to the low permeability of the rocks that trap the gas. A controversial and debated methodology to extract unconventional gas is "hydraulic fracturing," which sometimes is called "stimulation" or in short "fracing" or "fracking" (EU Parliament—DG for Internal Policies, 2011).

A hydraulic fracture is formed by pumping fluid at high pressure into the wellbore at a rate sufficient to increase pressure down hole to exceed that of the fracture pressure gradient of the surrounding rock. The fracturing fluids are commonly made up of water (over 90%) and chemical additives (Wang et al., 2014).

After creating the fracture, the operators prevent it from closing by introducing a proppant that keeps the fracture opened when the injection of fluid is stopped and pressure reduced. When the fracture is completed, the fluid reflows to the surface including the gas (EPA Ireland, 2012).

There is a large debate regarding hydraulic fracturing, that is supposed to have a relevant environmental impact due to the consumption of land, water and contamination from chemical additives. Moreover, in some cases, it is assumed that fracturing may induce earthquakes (EPA Ireland, 2012).

Despite these concerns, in the United States, the industry of unconventional gas continues to develop, whereas in the EU the situation is much more stable and the concerns linked to the environmental and security risks seem to prevail, also considering an ambiguous regulatory framework that has a number of gaps (EU Parliament—DG for Internal Policies, 2011).

10.3 Organization of Natural Gas Industry

Due to its complexity and the heterogeneity of the regulatory framework of the different countries worldwide, natural gas industry has an organization that differs, also substantially, from country to country, but some elements of the value chain remain the same, irrespective of the specific contexts.

Natural gas industry, as well as oil industry, can be divided in three main segments, namely, upstream, midstream, and downstream, each focused on specific aspects of the production and commercial process, as shown in Figure 10.1:

1. Upstream: it mainly represents the exploration and production phases. It involves the active searching for underground and underwater, conventional or unconventional, reserves of natural gas, as well as oil. Most of the profits of the oil and gas industry are determined in this phase. The upstream phase can be divided in six steps:

   a. Acquisition of the rights for exploration from the reserve holder

   b. Performing the surveys to find the reserves

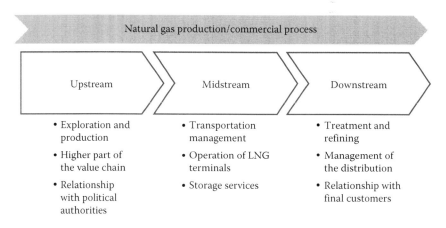

FIGURE 10.1
Fundamental phases of the natural gas production/commercial process.

    c. Execution of preliminary drilling tests to check the source

    d. Accomplishment of checks to determine the commercial viability of source

    e. Large-scale production (extraction) of natural gas (or oil)

    f. Payments to the reserve holders in the form of royalties or production sharing arrangements (PSAs)

2. Midstream: it represents the second phase of the production. It consists in the transportation of crude or refined oil and gas products via pipelines, tankers, trains, and so on. The final destinations are refineries or treatment plants, where the downstream phase begins. In the case of natural gas, there is also the operation of liquefaction and regasification terminals, which allow transforming natural gas into liquid phase and vice versa, in order to allow an easier transportation of relevant quantities. Also, storage services are typically included in the midstream phase.

3. Downstream: it is the final stage of the process. It is devoted to the refining, treatment, and purifying of oil and natural gas, as well as it includes all the efforts made to market and distribute natural gas (or oil) to final customers.

A company operating in the field of oil and gas is said to be "vertically integrated" if it covers all the phases of production, from upstream to the downstream.

In some countries, in order to support the competitiveness among the companies involved in the business, there are limits to the degree of vertical integration. This is supposed to prevent the formation of "barriers to entry" and dominant market positions.

10.4 European Regulatory Framework

During the 1990s, a huge debate was opened regarding the increase of effectiveness of the European gas industry. The main driver of the discussion was connected with the idea of supporting the free access of third parties to the gas network and the privatization and liberalization of the sector, which was, until then, dominated by vertically integrated state-owned companies.

This kind of organization was very common in Europe. It consisted in the presence of a national "oil company" in charge to supply fossil fuel, usually oil and gas, to the country.

This company often incorporated all activities of the fuel value chain, from the upstream up to the distribution to final customers. Of course, such a model limited the concurrency, because the vertical integrated operator could benefit from a favored position with regard to the distribution network and, consequently, the possible customer base (Bianco et al., 2015).

The main aim of this renovation process was to create a more convenient market for final users by breaking national monopolies and by creating a free market based on the concurrency of the operators.

The optimal target of this process would have been the creation of a single European market, with the price set by the interaction of supply and demand, namely, based on the "clearing price" mechanism.

A first step in this direction was achieved with the release of the 98/30/EC directive, also known as the "first gas directive," which aimed at creating a common framework for the EU gas market.

This directive presented for the first time the principle that consumers could freely choose their suppliers and established some basic rules for the settling of a European competitive market.

For the first time, there was the introduction of competition in the natural gas market, which was characterized by strong national monopoly, and "gas to gas" competition was also mentioned for the first time. "Gas to gas" competition refers to the concurrency of the gas supplies of the different operators working in a free market.

These changes were supposed to optimize the efficiency of the natural gas industry and to guarantee better supply conditions for final users.

The competition was gradually introduced firstly allowing power plants and large industrial users to freely choose their suppliers and subsequently opening the market also to the small consumers.

A fundamental issue for all the network industries (e.g., water, electricity, TLC) is represented by the management of the network, which can be seen as a "natural monopoly," because the infrastructure has to be managed as a whole and it cannot be split in smaller parts. Therefore, only one (or a few at maximum) subject can be in charge of managing it.

Natural gas grid is usually characterized by transmission and distribution networks; therefore, there is the necessity of a regulatory framework in order to guarantee the access to the third parties.

The directive of 1998 forces incumbent operators to guarantee third party access (TPA) to private operators that want to operate in the natural gas sector. To stress this aspect, the directive established the principle of the separation of the activities of the incumbents to ensure a transparent and nondiscriminatory access for third parties to the existing infrastructure.

The 98/30/EC directive represents a milestone, because it was the first clear step, at least from the "conceptual" point of view, toward the establishment of a free market, but its practical implementation in the different member states was very limited.

The inadequate implementation of the directive was due to the lack of specific prescriptions for the member states on how to implement the new market mechanisms; therefore, a lot was left to their willingness toward the implementation of a real liberalization process. In many cases, it was refused.

On the other hand, for EU authorities, the liberalization and creation of an integrated European gas market represented an important strategic goal; therefore, to support this position, the directive 2003/55/EC, also known as "second gas directive," was released in 2003 and it replaced the 98/30/EC directive.

This new directive had the objective to enforce the concepts already expressed in the 98/30/EC directive and to set a clear work plan, with mandatory milestones, to implement the liberalization process.

An example of the acceleration requested by the EU with the 2003/55/EC directive is represented by the mandatory establishment of an independent regulatory authority for TPA, which should monitor the market in order to avoid the presence of market concentration and the exertion of market power, especially from the former monopolists or incumbent operators. All this was only due on "voluntary base" according to the 98/30/EC directive.

The regulatory authority is also in charge of the ex ante approval of the access tariffs to the distribution and transmission pipelines. Exceptions to the TPA principle can be made in the case of new projects for important pipeline infrastructures, in order to stimulate the investments in the sector.

Furthermore, the directive also prescribed the legal separation of the transmission system operator (TSO) from all the other activities of vertically integrated operators. This aspect was further emphasized in the "third gas directive."

In 2009, the "third gas directive," 2009/73/EC, was issued and it represents the European gas legislation in force at present.

The directive 2009/73/EC is of fundamental relevance especially for the issue of the "unbundling," namely, the legal separation of the TSO from all the other activities of vertically integrated operators. To this aim, it proposes three possible models, namely, ownership unbundling (OU), independent system operator (ISO), and independent transmission operator (ITO). More details on these arrangements are reported in Table 10.1 (Bianco et al., 2014).

The directive furnishes a detailed framework with three specific, different options for the member states, which have to choose and implement the scheme more suitable for them.

In conclusion, as highlighted in Figure 10.2, it can be said that EU gas directives focus on three main issues (Bianco et al., 2014):

1. Unbundling of transport and other activities
2. Regulated third party access
3. Concept of "eligible customer"

The first issue is aimed at breaking down of vertical integrated companies, typically former state monopolists. To this objective, the directives establish the separation of companies dealing with the "raw materials" (producers, importers, wholesalers, retailers, etc.) and companies furnishing

TABLE 10.1

Description of Unbundling Models according to the Directive 2009/73/EC

Definition	Description
Ownership unbundling (OU)	A new company that owns and manages the transport network is created. This company results to be totally independent by the vertically integrated companies operating in the exploration, production, and retail business. In 2009/73/EC, OU is indicated as the most effective way to promote investments in infrastructure in a nondiscriminatory way, fair access to the network for new entrants, and transparency in the market.
Independent system operator (ISO)	Vertically integrated company maintains the ownership of the transport network, but its management is in charge of an independent company.
Independent transmission operator (ITO)	Vertically integrated company maintains the ownership of the transport network and the control of the company in charge of its management, but it must guarantee its independence. The independence of the transmission operator is assessed by controls of the national authorities.

Source: Bianco, V. et al., *Appl. Energy*, 113, 392, 2014.

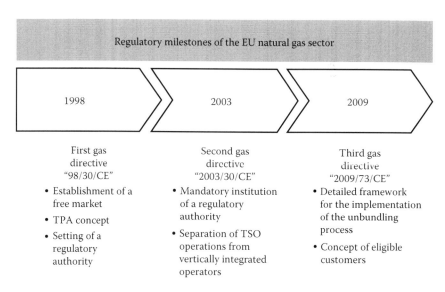

FIGURE 10.2
Development of the regulatory framework.

infrastructures and services (transporters, LNG plant operators, storage, etc.) to the system.

The second point is connected with the definition of a clear regulation to ensure the access to the transportation network to all the operators, in order to establish a fair competitive environment.

Finally, the third point focuses on the right of the customers to freely choose their natural gas supplier. In this way, it is supposed to ensure more convenient supply conditions for the final customers.

Thanks to these legislative changes, the European gas market was completely reshaped in the last 15 years and the level of concurrency was noticeably enhanced, as highlighted by the relevant number of wholesale and distribution companies nowadays active on the market (Bianco et al., 2015).

The future regulations currently under discussion aim at establishing liquid natural gas trading hubs, with the ambition to obtain a European reference price on the basis of the transactions, namely, the interaction between demand and offer, executed on the hubs (Stern, 2012).

10.5 Supply and Demand Balance in Europe

10.5.1 Analysis of the Consumption

The modern history of natural gas in Europe began in 1959 with the discovery of the Groningen field in the Netherlands, followed a few years later by the first discoveries in the UK sector of the North Sea. This was followed by equally substantial discoveries of gas in the Norwegian sector starting in the 1970s. But while the United Kingdom had a huge domestic market, Norway did not and created a huge export business with a number of pipelines delivering gas to both continental Europe and the United Kingdom (Stern, 2012).

Today, natural gas is used for a wide range of activities in everyday life, ranging from buildings heating to industrial processes. Because of this large variability within the natural gas consumers, they are usually grouped in consuming categories, in order to have a more coherent picture regarding the consumption of natural gas. Three consuming categories are commonly distinguished: residential/commercial, industrial, and power generators.

According to the economic theory of "demand and offer," the level of consumption depends on the price of gas, which is also different in the different areas of consumption. As a consequence of this, a reduction of consumption when the price increases and vice versa is noted. On the other hand, this implies that customers have a convenient and feasible fuel switching opportunity in a short time, but this is often not feasible, especially for some specific classes of consumers.

Residential users may think to switch from natural gas to fuel oil, electricity, or renewables to produce hot sanitary water, but to do this, the immediate availability of an alternative technology is necessary and this is usually not possible. Therefore, this option could be seen as a "long run" opportunity rather than a "short run" move. A similar consideration can be done in the commercial sector, in particular for small shops.

Industrial users or power producers are in a different situation. Most of them are provided with different kinds of equipment; therefore, they have more chances to pursue fuel switching strategies.

For example, a large power operator may decide whether to operate a natural gas or a coal power station belonging to its power plant's fleet.

In general, it can be said that residential and commercial gas demand is "less price elastic" than industrial and electric utilities' demand. In other words, residential and commercial natural gas demand is less sensitive to price changes, because these users have limited opportunities to use other sources of energy. On the other hand, industrial and power generation users have much more opportunities to diversify their energy sources; therefore, they are more "reactive" to the changes of price.

Residential and commercial customers are, in general, defined as "captive," due to their difficulties in reacting to the price signals; therefore, their consumption pattern is smoother with respect to other consumption categories.

Figure 10.3a reports the trend of energy consumption in EU15* from 1965 up to 2014 by showing the contribution of the different sources.

The figure shows that before 1970, energy consumption was dominated by oil and coal, whereas after that time the situation became more dynamic. This was due to the energy crisis of 1973, when there was the so-called OPEC embargo, which caused a sudden increase in oil prices. Since then, all the

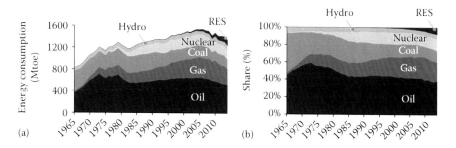

FIGURE 10.3
Historical trend of energy consumption in EU15 (a) and shares of different energy sources (b). (From BP, BP Statistical review of world energy, 2016, http://www.bp.com/en/global/corporate/energy-economics/statistical-review-of-world-energy.html. Accessed December 30, 2016.)

---

* EU15 includes Austria, Belgium, Denmark, Finland, France, Germany, Greece, Ireland, Italy, Luxemburg, the Netherlands, Portugal, Spain, Sweden, and United Kingdom.

policy makers understood the necessity to promote sources of energy other than oil, which is controlled by a restricted number of countries that can impose their rules.

Natural gas was immediately seen as a good substitute for oil and its consumption started to increase at impressive rates, whereas oil consumption began to decrease. From 1965 until 1980, gas consumption grew at the impressive average rate of 17%/year.

During the 1980s, natural gas consolidated its position as one of the fundamental sources of primary energy. In those years, also a relevant development of nuclear generation was observed, whereas the market shares of oil and coal decreased (Figure 10.3b). In the period 1980–1990, natural gas consumption grew at an average pace of 2%/year.

During the 1990s, a relevant increase of natural gas consumption was detected, with an average growth of 4%/year up to 2000, and this resulted in the establishment of natural gas as the second source of primary energy in EU15 after oil. On the other hand, a strong decrease of coal consumption was observed.

This noticeable increase of natural gas consumption could be ascribed to its large utilization in thermal power plants in substitution of coal, which has a much heavier environmental impact (e.g., in terms of CO_2 emissions) and requires the utilization of more expensive and sophisticated power plants in order to be compliant with EU regulations on environmental issues.

This tendency was also supported by the liberalization process of power generation in the EU during the 1990s, which opened the power generation sector to the free market by abolishing the former vertical integrated national monopolies.

This process allowed many investors and companies to start their power generation business in many EU countries and, often, the first move to enter the market was the implementation of a simple and efficient power plant, namely, a CCGT plant. Therefore, many operators built CCGT plants all over Europe, and this contributed to a sharp increase of natural gas consumption in the EU and in particular in the power sector.

As a consequence of this continuous growth, during the 2000s, natural gas became the second source of primary energy in the EU and it gained a relevant strategic importance for many countries.

In the period 2000–2008, its consumption continued to increase at an average pace of 1.5%/year, due to the construction of new gas power plants, also in light of the high outlook of carbon prices in the EU-ETS market, and its massive utilization in all sectors of activity.

This trend changed in the period 2008–2014, where an average decrease of 4.1% in consumption was registered. Only in 2010 there was a strong increase of consumption, up to the level of the years 2007–2008, but this gain cannot be explained by a recover in industrial or power production, and it is probably due to space heating in residential and commercial sectors, because of the cold winter of that year (Honorè, 2011).

This decrease is due to different factors, namely, the effect of economic crisis that determined a decrease in energy consumption, as shown in Figure 10.3a, and, in particular, of gas and electricity consumption. Less electricity consumption means less gas used in thermal power plants, which often represents the marginal unit in the EU power markets.

Moreover, as shown in Figure 10.3a, in that period there was the deployment of a large amount of renewable power plants that displaced natural gas power plants from the merit order and gained market shares. It should be also mentioned that in 2011–2012 there was a relevant reduction of coal prices; therefore, the power operators with coal plants in their portfolios switched the generation on the coal units, which proved to be more convenient (EIA, 2013).

Finally, the effect of energy efficiency policies put in place by the EU in order to limit primary energy consumption in the buildings and industrial sectors should also be taken into account. All these factors negatively affected the consumption of natural gas in the EU, which for the first time in history is declining.

Table 10.2 reports the natural gas consumption in EU15 and it can be observed that the consumption is not homogeneous. In fact, Germany, Italy, and the United Kingdom account for a large share of the total consumption. Specifically, Italy and the United Kingdom use a relevant amount of natural gas for power generation, whereas in Germany the consumption is concentrated in other sectors.

In conclusion, it can be said that recently the gas demand in the EU has weakened mainly as a consequence of the economic crisis but is expected to rise again along with the economic recovery, which means that short-term fluctuations are not likely to have a decisive impact on the long-term development of the EU's gas demand (Makinen, 2010). However, the view on the future consumption of natural gas in the EU is also influenced by other factors, such as the future climate policies, support to energy efficiency policies, promotion of renewables, and so on.

The implementation of these measures may determine a modest growth of natural gas consumption in the years to follow.

10.5.2 Analysis of the Supply

Natural gas considerably differs from other sources of fossil fuels in terms of supply. In particular, its transportation and storage are much more difficult in comparison with other fuels, such as oil and coal. In fact, both oil and coal can be easily shipped and delivered all over the world with a high degree of flexibility in terms of quantities to deliver and locations to reach.

On the other hand, natural gas delivery is rather inflexible. Most of natural gas is delivered by pipelines, which connect the extraction fields with the consumer; therefore, there is a close interdependence between consumers and suppliers or sellers and buyers (Makinen, 2010).

TABLE 10.2

Natural Gas Consumption for Each Country of EU15 in Billions of Cubic Meters (bcm)

Billions of Cubic Meters (bcm)	1995	2000	2005	2010	2011	2012	2013	2014
Austria	7.60	7.77	9.73	9.80	9.25	8.85	8.36	7.61
Belgium	12.66	15.94	17.57	20.27	18.04	17.08	17.17	15.08
Denmark	3.78	5.31	5.25	5.28	4.43	4.14	3.95	3.37
Finland	3.39	4.08	4.29	4.58	4.01	3.57	3.41	3.00
France	35.28	42.66	48.93	50.74	44.18	45.61	46.52	38.50
Germany	80.26	85.70	92.77	90.38	83.00	80.86	87.19	76.47
Greece	0.05	2.03	2.81	3.86	4.74	4.49	3.86	2.96
Ireland	2.78	4.10	4.14	5.60	4.91	4.78	4.61	4.45
Italy	53.13	69.11	84.27	81.18	76.12	73.19	68.45	60.48
Luxemburg	0.66	0.80	1.40	1.43	1.23	1.25	1.06	1.01
Netherlands	41.14	41.74	42.13	46.76	40.79	39.06	39.42	33.91
Portugal	—	2.43	4.47	5.35	5.32	4.80	4.48	4.14
Spain	9.21	18.15	35.59	37.13	34.58	33.69	31.20	28.23
Sweden	0.90	0.93	1.00	1.75	1.38	1.20	1.14	0.95
United Kingdom	77.67	104.24	101.94	101.15	83.73	79.09	78.34	71.25

Source: ENI, World oil and gas review 2016, 2016, https://www.eni.com/en_IT/company/fuel-cafe/world-oil-gas-review-eng.page. Accessed December 30, 2016.

This connection is enforced also by the typology of commercial transactions, often based on bilateral long-term agreements, usually in the range of 10–30 years, between buyer and supplier.

As discussed in the following, these contracts are "oil-indexed" and include the "take or pay" (TOP) clause, according to which customers are required to pay for a certain volume of gas even though they do not take the delivery of all of it. The TOP clause has the function to mitigate the "volume risk" for the supplier (i.e., the extraction of a volume of gas that remains unsold).

Natural gas market largely differs from oil market and it is developed on a regional basis, also in light of the fact that it cannot be easily transported. This fact has determined a close relationship between supply and consuming countries of the different regions. Only in the last years, with a larger diffusion of liquefied natural gas (LNG), the market is becoming more global and new suppliers are emerging on the market.

In general, most of the natural gas is supplied to Europe by means of pipelines. In 2014, this share reached ~87%, whereas the remaining part is supplied by LNG. The supply via pipelines is dominated by Norway, Russia, the Netherlands, and, to a lesser extent, Algeria, which, together, covered ~90% of the pipeline supply in 2014. Other suppliers are the United Kingdom, Libya, and other European countries.

The two largest EU suppliers are undoubtedly Norway and the Russian Federation. Between these, Norway is a member of the European Economic Area and therefore part of relevant EU regulatory frameworks, whereas Russia is the only external supplier enjoying a significant or even dominant position in both Western and Eastern European countries (Goldthau, 2013).

Russia gained importance in Western European gas supplies since West Germany signed its first long-term gas contracts with the USSR in the early 1970s. In the former communist EU member states, Russia retained its role as a dominant gas supplier after the fall of the Iron Curtain. Here, dependency rates are up to 100% of some countries' imports. Europe's high dependence on foreign sources of natural gas, notably from Russia, has caused security concerns among observers (Goldthau, 2013).

As for LNG, Qatar is the main supplier, with a market share of approximately 52% in 2014, followed by Algeria and Nigeria. Other suppliers are Trinidad and Tobago, Perù, and Oman.

Figure 10.4 reports the supply of natural gas to EU15 for country of origin in 2014. Figure 10.4a highlights that a relevant amount of gas comes from Europe (i.e., indigenous production), but European reserves are supposed to decline in the next years unless new basins are discovered; therefore, this share is expected to decrease in the future.

As shown in Figure 10.4a and b, the countries of origin of natural gas are quite various and a considerable share of the supply comes from countries or regions that are unstable from the political point of view. This raises many concerns related to the security of supply, because it exposes European

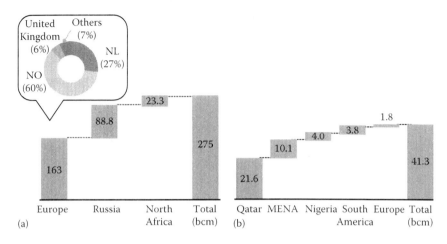

FIGURE 10.4
Supply of natural gas by country of origin by pipelines (a) and LNG (b) in 2012. (From BP, BP Statistical review of world energy, 2016, http://www.bp.com/en/global/corporate/energy-economics/statistical-review-of-world-energy.html. Accessed December 30, 2016.)

consumers to the effect of extra-European disputes, as was the case of the Russian–Ukrainian crisis of the winter 2009.

The 2014 tensions between the Russian Federation and Ukraine reignited European concerns about the security of its natural gas supply. Civil war in Ukraine and the sanction policies of the West and Russia have led to fears that Russian natural gas supplies will be interrupted not just to Ukraine but also the EU. At first glance, the dispute over natural gas prices and potential interruptions to supply was comparable to 2006 and 2009, although the situation is more severe with an actual looming war between Russia and Ukraine. On the other hand, the EU seems better prepared for any disruption of Russian supply with respect to the past (Richter and Holz, 2015).

It should also be mentioned that the EU has a pivotal role as the main customer of the Russian Federation, even though the Asian market is an attractive alternative for Russia, with good prospects (Paltsev, 2014) in the short run. On the other hand, actual trade flows are limited due to a lack of production and transportation infrastructure in East Russia. Thus, different years and huge investments are required before Asian markets become a real attractive alternative for Russia.

One of the commonly acknowledged weaknesses related to natural gas as an energy source is the lack of adequate "shock absorbers" that allow the supply system to respond to sudden unexpected increases in demand or loss of supply.

In general, natural gas production is closely related to the demand. If there is a decrease of the demand, production is reduced correspondingly, whereas production is increased only if "pulled" by the demand.

The reason for this mechanism is due to the difficulty in storing natural gas; in fact, storage facilities are rare and quite expensive to build. Moreover, because most of natural gas is supplied through pipelines, it is difficult to find alternative markets to reach. The inadequacy of storage capacity is a great problem for several EU member states and reduces their abilities to react in the event of an energy crisis (Makinen, 2010).

10.5.3 Development of Infrastructure

Because of the difficulty in transporting natural gas from the extraction to the consuming locations, infrastructures play a fundamental role in the natural gas industry.

As observed by Dieckhöner et al. (2013), the European gas market is confronted with significant challenges over the next years. Within the borders of the EU, natural gas production is declining due to limited reserves. This especially affects today's largest gas-producing countries in the EU, the United Kingdom and the Netherlands.

In order to import the expected increasing natural gas volumes, an expansion of the import capacity for natural gas into the EU will be necessary. These additional volumes delivered at the EU border by pipelines or as LNG will also affect the gas flows within the EU, as this natural gas has to be transported to final consumers.

To receive these new volumes, it will be necessary to expand cross-border capacities, in terms of pipelines, LNG terminals, and storages. Storages are important to allow a possible decoupling between demand and supply, especially if many pipeline connections are planned (Dieckhöner et al., 2013).

Another important issue to take into account in the development of new infrastructure is the "security of supply." For example, during the 2009 Russian–Ukrainian gas conflict or the Libya crisis of 2011, some EU consumers experienced short-term supply disruptions.

Western and Central Europe avoided disruptions due to diversified supply portfolios and transport routes, sufficient natural gas storage, and high physical market integration. The high level of integration allowed the transportation of gas volumes against the normally prevailing flow directions and, thus, to supply countries that are dependent on the Ukrainian route (e.g., Hungary).

The Russian–Ukrainian crisis gave two important lessons to the EU (Dieckhöner et al., 2013):

1. Security of supply is not the same within the EU.
2. A physical market integration can significantly improve security of supply and mitigate the danger of supply disruptions.

To these ends, the Third Energy Package of EU legislation on the internal electricity and gas markets promotes the facilitation of cross-border trade in energy, cross-border collaboration and investment, and the enhancement of increased solidarity among the EU countries.

After more than a decade of announcement and planning of new infrastructures to connect the producing gas fields in Russia, Central Asia, and the Middle East, three new pipelines are under construction or preliminary work on them has begun. These pipelines are designed to supply Turkey, Bulgaria, Greece, and Italy with gas from Azerbaijan.

These pipelines follow the inauguration of the Nord Stream pipeline in late 2011. Since the beginning of the operations of Nord Stream, Russian exports of natural gas via the Ukraine have diminished from 65% of total Russian natural gas exports to Europe in 2010 to about 50% in 2013.

The direct link to a West European importer has stopped the long-term reduction trend of the Russian share in total EU natural gas imports (Richter and Holz, 2015).

The South Caucasus Pipeline Expansion will transport gas from Azerbaijan to the border with Turkey and Georgia. The Trans Anatolian Natural Gas Pipeline will transport gas from the Turkish–Georgian border to the Turkish–Greek border. Finally, the Trans Adriatic Pipeline will transport natural gas from Greece to Italy.

In the past years, there were several proposals to build pipelines through Turkey to supply the Turkish market, which is in expansion, and the EU markets. These infrastructures had a twofold objective, namely, to diversify the routes of supply (e.g., for Russian producers reducing transit risk by bypassing Ukraine) and the supply sources (e.g., for EU customers to obtain supply from Central Asian countries). For various reasons, these projects were abandoned or are on hold.

The Nabucco pipeline project was planned to supply southeastern Europe with natural gas and to connect to the EU natural gas network in Austria, but it was abandoned in 2013.

The South Stream pipeline, aimed at transporting natural gas from Russia to Bulgaria and Eastern Europe, was close to beginning construction, but there were delays connected with disagreements with the EU on the market legislations. Discussions to restart the project stalled after Russia's involvement in Crimea, and in December 2014 Russian President Putin canceled the South Stream project (EIA, 2015). At the same time, there was the announcement to propose a new pipeline called Turkish Stream, which would have transported natural gas from Russia to Turkey, across the Black Sea, and then onshore to Greece and from there to Austria through the Balkans.

Russia and Turkey were not able to reach an agreement, and the future of this infrastructure is currently uncertain (EIA, 2015). Figure 10.5 reports a map with LNG and pipeline import infrastructures from external suppliers updated to 2014 (Richter and Holz, 2015).

It should also be mentioned that the current market context with low demand and low prices is not particularly favorable to the planning and construction of new infrastructures.

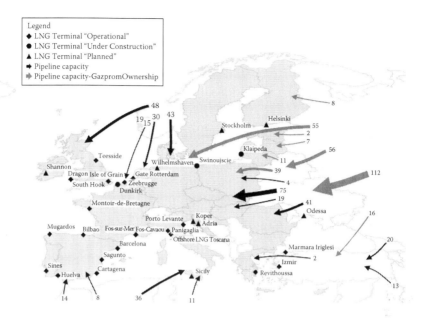

FIGURE 10.5

Map reporting the most relevant European infrastructures. (From Richter, P.M. and Holz, F., *Energy Policy*, 80, 177, 2015.)

Other possibilities for gas import in the EU involve the supply from Iran and Iraq through Turkey.

The lifting of sanctions on Iran would allow European countries to import gas from Iran. Although Iran already exports natural gas to Turkey, it has long had plans to export larger volumes of natural gas through Turkey to Europe. However, other hurdles would remain, including agreeing on a natural gas price and meeting Iran's growing domestic demands for natural gas, especially for enhanced oil recovery, power generation, and winter heating (EIA, 2015).

Another project under discussion is the GALSI pipeline, which should connect Algeria and Italy.

In parallel with pipeline plans, also projects based on LNG infrastructures are under development. LNG allows adding more flexibility to the supply of natural gas, because more suppliers can be involved from all over the world. This reduces the dependence of the EU on supply from pipelines that depend on countries such as Russia and Algeria.

The possibility to access the LNG market gives to the EU buyers a very powerful instrument in the negotiation or renegotiation of long-term supply agreement for pipeline gas, because they can leverage on the possibility to sign obtain competitive LNG provisions.

LNG is supposed to increase the competition on the natural gas market and it can be seen as an element of "globalization" of the gas market, which, in general, has a regional dimension.

At present, 23 LNG terminals exist in Europe for a gasification capacity of ~190 bcm (Richter and Holz, 2015), other terminals are under construction/committing for an additional capacity of 35 bcm, and 32 LNG terminals are under development/planning for a further potential capacity greater than 160 bcm (Renier, 2013).

In principle, LNG terminals are easier to develop if compared to pipelines, because they are the result of a "simple industrial project," and fewer "geopolitical" issues are involved.

The development of LNG terminals permits the importing of large quantities of natural gas from a large variety of suppliers. In particular, after the development of unconventional gas, new suppliers, such as the United States and Canada, are available on the market.

Apart from the commercial point of view, the possibility to have more suppliers, some of them with a reliable political context, is fundamental in order to ensure the security of supply, which represents the focus of the current EU energy strategy.

10.6 Natural Gas Pricing

10.6.1 Market Context

If a comparison between oil and gas literature is performed, it can be seen that studies focused on oil are much more numerous with respect to those addressing natural gas market, despite the fact that gas is approaching 25% of the global primary energy consumption.

In the majority of countries outside North America, international gas prices are not transparent and accurate public domain data are very difficult to obtain. This may not have been a great problem when the fuel comprised only a few percentage points of energy balances but as gas has become more important, so has the way in which it is priced (Stern, 2014).

The vast majority of international gas trade outside North America is still conducted on the basis of 10–30 years' contracts with complex price clauses, based on private negotiations (Stern, 2014). Therefore, the final price is influenced by the political relationships between supplier and buyer, negotiation power between them, and by other features not necessarily linked to the fundamentals of gas industry.

On the contrary, in the United States, natural gas transactions are based on the Henry Hub price, which is public; thus, the market is much more transparent and the price is closely linked to the market fundamentals.

A factor that may affect both the demand and the supply, and thus the price, of gas is the price of oil. The price of oil is relevant because of substitution properties of gas and oil.

In principle, substitution is primarily relevant in the electricity generation and the heavy industry. If the price of oil rises, burning gas becomes relatively cheaper, increasing the demand for gas, which in turn results in an upward pressure on the gas price (Hulshof et al., 2016).

However, this is more a theoretical statement, rather than a practical evidence; in fact, Stern (2012, 2014) argues that short-run fuel switching is hardly relevant anymore in West Europe, because oil has virtually disappeared in most stationary energy sectors, maintenance of dual-fuel burners is expensive, and because of tight environmental standards as well as the inefficiency of using oil in new gas-burning technologies. In other words, oil has been almost completely substituted by natural gas and in large power or industrial plants oil is basically used only in case of emergency, rather than for fuel switching purposes due to possibly favorable market conditions.

Apart from the lack of short-run substitutability, there are other relevant reasons why oil and gas should have different price evolutions. The value chains of oil and gas are substantially different; in fact, transportation, production, and storage facilities have completely different characteristics, which determine a totally different cost structure.

Moreover, oil and gas markets developed with diverse features. Oil market is based on the world context of supply and consumption. Although regional differences between types of crude oil exist, such as Brent and West Texas Intermediate, their prices are usually correlated. On the other hand, natural gas price is much more subjected to regional dynamics of demand and supply. Regional prices can move in very different directions and on diverse levels, as can be seen from the recent divergence in European and North American gas prices (Hulshof et al., 2016).

This is due to the fact that the market is less flexible due to the difficulties in transporting natural gas. For example, if there is a pipeline connection, demand and supply are "rigidly" connected, because there is not a wide range of alternative markets for both sellers and buyers, and therefore the price has been usually determined on the basis of long-term bilateral contracts. This market structure has characterized the European context from the 1960s to the present.

10.6.2 Oil-Linked Formulas

During the 1960s, natural gas begun to be consumed in significant quantities and to be seen as the ideal substitute for oil. Natural gas market developed on a regional basis, in opposition to the oil market that developed on a global scale, due to the difficulty to supply natural gas all over the world. Therefore, the natural gas market assumed different configurations in different geographic areas; in particular in the United States and the United Kingdom, price level was determined on hub markets, while in continental Europe the long-term oil-indexed gas price was adopted, and in the Soviet area there was a regulated price model (Stern, 2012).

Accordingly, European consuming countries stipulated long-term supply agreement with producing countries, namely, Norway, Russia, the Netherlands, and Algeria.

Large pipelines were built to connect supply with consuming countries and the companies that constructed and managed such large infrastructures needed to get an adequate reward. In light of this, long-term contracts, with a duration of up to 30 years, were signed to secure the takeoff of an adequate quantity of natural gas in order to guarantee, at least, the payback of the investments.

These agreements were based on economic and market fundamentals. Economic fundamentals refer to the cost of developing and delivering domestic or imported gas to end users. Market fundamentals refer to the price of gas, compared with the price of market substitutes (Stern, 2014).

At that time, there were no liquid markets where gas price could be determined by the interaction of supply and demand; therefore, the idea of linking the price of gas to its closest substitute, oil, grew.

The logic of this mechanism was based on the fact that natural gas represented a good substitute for heating oil and, moreover, a well-established market (i.e., reference) for oil was already developed.

According to this, long-term agreements between producing and consuming countries were signed and their prices were based on bilateral negotiations and indexed to oil.

The logic of the division of risk inherent in these contracts was as follows (Stern, 2014):

- The exporter assumed the price risk, that is, the risk that the price, however determined, would be sufficient to remunerate the investment in production and transportation of gas to the border of the importing country.
- The importer assumed the volume risk that a large enough market would be developed in order to honor the volume commitments in the contract. This risk was formalized by means of a "take or pay" clause, which imposed to the buyer to withdraw a fixed amount of gas volume per year and if it was not able, it should pay for it in any case.

The formulas reported in these contracts assumed that gas price is affected by a number of parameters, namely, prices of competing fuels, GDP growth rates, inflation and taxation, industrial structure, environmental regulations, and a range of other country- or region-specific conditions that change over time. Therefore, the cost of gas supply varies according to the variation of the parameters included in the formula.

In these contracts, prices are generally adjusted quarterly, based on an average of (mainly) oil prices in the preceding 6–9 months, with a lag of 3 months. Thus, the buyer pays a price in the first quarter of a year related to an average of oil prices in the first two or three quarters of the previous year (Stern, 2014).

The long-term supply contracts often include the possibility to perform a "price review," which allows modifying the structure of the formula upon request of seller/buyer or in the presence of unexpected events (e.g., relevant changes to the gas market, significant modification of the geopolitical context). Usually, a price review is set every 3 years, but in most cases the possible changes to the structure of the formula are very limited.

The consolidation of the oil–gas price relationship in Europe has resulted in the fact that people are used to the circumstance that gas prices rise because of changes in the oil markets. Many large industrial consumers take oil-indexed gas pricing for granted as they are convinced by suppliers that it is safer.

The role of long-term contracts in the development of the natural gas market can be considered controversial. As observed by Dilaver et al. (2013), on one hand, they can be seen as "entry barriers" for potential and more efficient suppliers, but, on the other hand, they are supposed to simplify the market entry by supporting long-term investments.

As a matter of fact, the management of long-term supply agreements demonstrated to be easier for both the sellers and buyers, with respect to a "free gas" market. The structure of these contracts guarantees the supply for a long period, so that the consumers could take their investment decisions (e.g., power generators), whereas the suppliers could easily manage the delivery, for example, by securing transport capacity for a long period, without the necessity to continuously participate in periodic capacity market auctions.

A main criticism of the contracts is that they are not connected with the economic fundamentals of the natural gas value chain, but based on the oil one, which is completely different. On the other hand, despite all the possible criticism, these agreements have dominated the European gas market for more than 40 years.

An example of a long-term oil-indexed formula can be expressed in the following form (Bianco et al., 2014):

$$P_m = P_0 + \gamma_1 \left(LFO_m - LFO_0 \right) + \gamma_2 \left(HFO_m - HFO_0 \right) \tag{10.1}$$

where

LFO_0 and HFO_0 are the starting price of light fuel oil and heavy fuel oil

LFO_m and HFO_m represent the price for the month m, which generally takes the average value of the previous 6–9 months

γ_1 and γ_2 are two coefficients taking into account the natural gas market segments competing with HFO and LFO, factors to share risks or rewards between sellers and buyers and technically converting factors to have homogeneous units of measures

Finally, P_m is the price paid by the buyers in the month "m," whereas P_0 depends on the starting price of natural gas when the supply agreement is signed and by the private negotiations between buyer and seller. Thus, the formula can be split into two contributions: P_0, substantially determined by market conditions at the time of the agreement plus private negotiations, and the second contribution linked to the oil spot market. In other words, P_0 represents the basis on which the formula is settled (e.g., the base price level), whereas the second part describes the evolution during the time according to the trend of the oil market.

As previously mentioned, the formula also includes a "take or pay" (TOP) clause in order to establish a minimum quantity of gas to withdraw and a "price review" clause, usually to be exerted every three years, to include slight modifications in the formula.

Obviously, these formulas are not publicly available and they often represent an industrial secret for both buyer and seller; therefore, there is a lack of data, which makes quite difficult any kind of analysis and specific evaluation of the European gas market (Stern, 2012).

10.6.3 Gas Hubs

Trading hubs can be seen as "points" in a natural gas pipeline where gas is exchanged between sellers and buyers. A hub can be physical or virtual. In a physical hub, natural gas is injected or withdrawn at a specific point of the pipeline, whereas a virtual hub covers a network area and the gas injected or withdrawn from this area is associated with the reference virtual hub. Table 10.3 reports a list of the main European hubs and their major features.

A physical hub offers the advantage to express a localized price signal, but it complicates the creation of a liquid market, because natural gas should be delivered and withdrawn at a specific point in the network. This necessitates that buyers and/or sellers should purchase the pipe transportation capacity to reach the physical hub. Therefore, this means that a physical hub needs a relevant infrastructure to connect it with the rest of the network in a way that the transactions could be easily performed, namely, that a liquid market can be created. This implies very high investment costs.

Virtual hubs are introduced in order to avoid these problems and to create a more flexible market. They have a number of entry and exit points; therefore, the market operators only have to be sure that gas is delivered to one of the inlets of the network or withdrawn from one of the outlets. In this way, market participants have more options to interact with the hubs, and this tends to increase the liquidity of the market. This structure simplifies the trading activities, because the different possibilities in terms of entry/exit points in the network give the opportunity to use different pipe routes to interact with the hub.

Since the entry and exit charges do not depend on the location of the virtual hub where natural gas is finally injected or withdrawn, all the gas can be taken or delivered at the same price within the network representing the hub. In light of this, the area connected with a virtual hub expresses a single

TABLE 10.3

Principal European Gas Hubs

Name	Description
National Balancing Point (NBP), UK	Started in the United Kingdom in 1996, it is the most liquid hub of Europe.
Zeebrugge, BE	It is the first trading hub of continental Europe and it started its trading activity in 2000.
Title Transfer Facility (TTF), NL	It was established as a virtual hub in 2003 for trading on Netherlands' national transmission grid. In 2005, it was opened to the international trading. It is the most liquid hub in continental Europe.
Punto di Scambio Virtuale (PSV), IT	It is the virtual hub of the Italian market, developed with the intention to allow more advantageous price to Italian consumers.
PEG, FR	Similar situation as Italian PSV.
Gaspool, DE	Developed in Germany in 2009, these two hubs are
NCG, DE	expanding rapidly. Their combined trading is larger than TTF.
CEGH, AT	CEGH started traded activity in 2005. It is an important hub to monitor, as it will play a role in introducing "gas to gas" competition and market pricing in Central and Eastern Europe. Its delivery point is Baumgarten, where pipelines originating in Russia diverge to supply gas to Austria as well as to Germany, Italy, and Hungary through transit pipelines.

price linked to the number of trading operations, that is, liquidity of the hubs, executed on the hubs.

Natural gas trading hubs have a twofold function: the first one is to balance demand and supply within a specific area. The interaction between demand and supply also determines a price (i.e., clearing price) for the delivery or withdrawal of gas on the hubs for a specific date.

The second function is that a trading hub serves as a source of physical flexibility to balance supply and demand. In fact, an excess of production can be sold on the hubs, as well as if some operators are "short" they can buy gas on the virtual hubs.

The problem with physical transactions on gas hubs (i.e., commercial operations that end with physical exchange of natural gas) is that simultaneously with the commercial operation, it is necessary to develop a corresponding "physical operation" consisting of buying the necessary pipeline capacity to transport natural gas from the hub to the location of delivery (which is usually external to the hub and even located in another country) or vice versa. This difficulty has limited the role of the hubs only to the physical exchange of residual quantities. Moreover, there is also an issue connected with the "flexibility" of the supply.

Inflexible supply of gas includes pipeline-contract gas up until TOP volumes, destination-inflexible LNG cargoes, and indigenous production that does not seem to respond to hub price signals in practice. While these tranches may have some flexibility (e.g., to allow for seasonality), they generally flow irrespective of the absolute level of hub prices and have therefore no primary impact. The flexible supply of gas consists of pipeline-contract gas between the TOP and maximum annual contracted volume, uncontracted pipeline import flexibility, and flexible LNG supply (Hulshof et al., 2016).

10.6.4 Hub Pricing

As given in the previous section, gas hubs are "physical" or "virtual" in location, where natural gas is exchanged among market operators; thus, there is a close interaction between demand and supply and the respective curves in a given moment can be obtained, as sketched in Figure 10.6a.

The clearing between supply and demand determines the price level expressed on the hub. In this way, the price of natural gas is only an expression of market fundamentals, namely, demand and supply levels, and it is not correlated to its potential substitutes.

From the side of the supply, a "gas to gas" competition is created, because a sort of "merit order" of the supply is determined and the level of the demand will set the market price (Figure 10.6b). This principle is economically justified as supply and demand of natural gas determine price.

To avoid abuse of market power (pushing prices up), competitiveness is assured by giving access to many players on demand and supply sides.

With the progress of gas market liberalization in Europe, gas systems moved from a monopolistic to a more fragmented environment. In the former, a single vertically integrated company managed most of the injections and withdrawals in the gas network. In the latter, instead, different agents cover a smaller share of the aggregate traded gas volumes, increasing the number of associated exchanges. A gas hub offers a way to clear individual positions, easing the need to balance physical injections and withdrawals. In turn, as liquidity develops, price signals become more reliable and a wholesale market offers a second source of gas provision in alternative to the traditional long-term contracts (Miriello and Polo, 2015).

On the other hand, it should be also mentioned that, although a large number of traders are active on the gas hubs, the supply to the gas market is concentrated because of the limited number of producers. The limited number of producers of gas within the European context raises some concerns on the degree of competitiveness of the gas market (Hulshof et al., 2016).

However, as a consequence of the numerous transactions, influenced by structural and random events, hub prices are much more volatile than long-term contract rates. For example, a more pronounced seasonality trend, due to climate conditions, can be observed with respect to long-term contracts

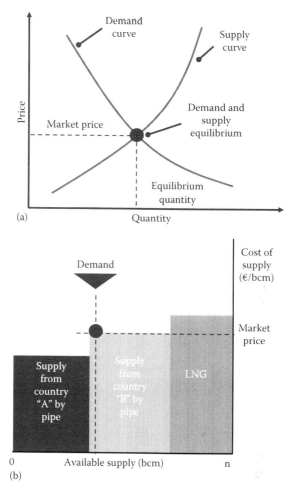

FIGURE 10.6

Natural gas market: sketch of supply and demand equilibrium (a); an example of natural gas merit order (b).

(Davoust, 2008). In fact, during the summer period, when heating systems are switched off, a relevant drop of the prices is detected, because the level of the demand is very low; on the other hand, there is an increase of the demand during the winter, with a consequent rise of the price. Therefore, an immediate reaction of the market to the changes of demand level can be observed and the price changes accordingly.

The hub trading mechanism tends to increase the elasticity of the demand with respect to the price (i.e., demand is more reactive to price signals), but in order to fully achieve this object, it is necessary to develop an adequate infrastructure.

In fact, once the gas is "financially" traded on the hub, then it is necessary to guarantee the corresponding physical delivery by means of pipeline networks and, to this end, it is necessary to have a transparent and efficient transportation and distribution capacity market to allow all the operators the possibility to deliver the gas to final customers. Therefore, there is the necessity to develop a parallel capacity market in order to sustain the growth of the gas market; this requires the close cooperation of the EU transmission system operators (TSOs) to guarantee the security and the correct balance of the network (Bianco et al., 2015).

10.6.5 Recent Developments

The development of gas hubs in Europe started around 2005, with the exception of NPB in the United Kingdom that begun its operations in the mid-1990s. Until 2007–2008, gas hubs of continental Europe did not show price levels significantly different from oil-indexed contracts and they had a limited liquidity.

As observed by Bianco et al. (2014), after 2008 this situation was radically transformed as a consequence of the general economic downturn and due to some specific events:

- Economic downturn caused a reduction in natural gas demand, with the consequence of increasing volumes on the market.
- Similar to natural gas, also electricity demand decreased and electricity generators tried to sell on the gas market part of the "take or pay" quota of their contracts, in order to reduce financial losses.
- Because of the strong development of unconventional gas extraction in the United States, a huge quantity of liquefied natural gas (LNG), originally directed in the United States, is diverted toward European and Asian markets.

In light of this context, power generators with large supply contracts in their portfolio to contain their financial losses tried to sell part of their TOP on the gas hubs. Similarly, gas suppliers, to face the drop in consumption, offered large quantities of gas on the European hubs, as also the LNG operators, who decided to adopt the same strategy to sell the volumes originally directed to the United States, where, in the meanwhile, shale gas sector had an unforeseeable growth.

The sum of these events determined a substantial increase of liquidity on the hubs, and, according to the general market law of equilibrium, when the offer increases, a drop in the prices is obtained. Therefore, prices on the hubs "decoupled" from oil-indexed contracts and set on much lower levels, as reported in Figure 10.7 (Stern, 2014). On the other hand, the price of the gas based on oil-indexed contracts continued to increase, because oil prices had an upward trend; thus, oil-indexed and hub-based gas prices showed

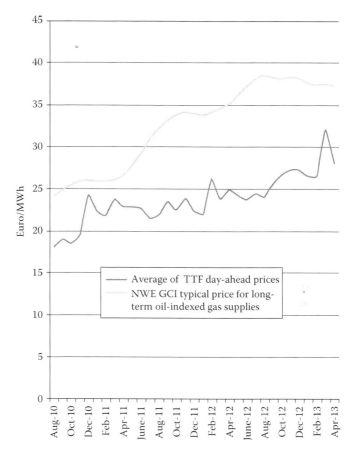

FIGURE 10.7
Monthly averages (August 2010–April 2013) of European long-term contracts and TTF spot prices. (From Stern, J., *Energy Policy*, 64, 43, 2014.)

an uncorrelated behavior. This situation was new and unexpected for the European natural gas market.

As observed by Stern (2012), from 2008 the usual commercial environment for European gas companies has been subjected to a number of new (and difficult to predict) forces, which have exacerbated the problems of reliance on the relatively rigid oil-linked price formulas in long-term contracts.

This situation caused the renegotiation of many oil-indexed contracts, because the buyers observed very low prices on the gas hubs if compared with the values of their contracts and therefore they claimed for supply conditions aligned with the market levels. Clearly, not all the negotiations had a successful ending and many arbitrage processes were also opened. This condition was typical of large and very large operators with huge supply contracts in portfolio.

On the other hand, small power operators and natural gas wholesalers tried to exploit this situation by buying natural gas volumes on the hubs by benefiting from the very low prices, therefore offering electricity and natural gas at very competitive prices.

This situation forced large operators to offer their power plants or gas quantities at prices based on the hub level, whereas their cost structure was much higher, based on the oil-indexed contract level. This strategy was implemented in order to avoid their displacement from the market, but it caused huge financial losses.

Stern (2012) highlighted that this situation generated a great confusion, because hub prices have been perceived as "more convenient" with respect to oil-indexed formulas, but the fallacy of this equation is rather evident, because there is the mistake of confusing price formation with price level. Thus, the assertion that when the gas supply/demand balance tightens, gas prices will "recouple" with oil prices, reflected this confusion.

A tight supply/demand balance will certainly result in higher prices, but there is not necessarily a relationship between the latter and oil-related price levels (Stern, 2012).

This is certainly true from the theoretical point of view, but from the practical point of view, it might be considered a "rule of thumb" that the oil-indexed gas formulas represent the highest limit for price levels. If gas price on the hubs should exceed oil-indexed formulas, it does not make any sense for large consumers to purchase on the hubs due to the complex issues connected with the physical delivery; on the contrary, it is more convenient to sign a long-term agreement to obtain a stable furniture, even though less flexible.

10.7 Conclusions

European natural gas sector is experiencing a period of radical changes, which were in act before the economic downturn. On the other hand, the economic crisis has exacerbated and accelerated these processes.

In particular, it is highlighted how the oversupply condition due to the low demand on both power and gas markets represented a "boost" for the gas hubs, which, in a short time, moved from low to high liquidity, becoming an important reference for the EU gas markets.

This situation has weakened the link between oil and gas markets, as auspicated by the EU directives. A hub-based gas market implies that natural gas price is connected with the dynamics of gas production and supply chain, rather than to the oil sector, which is considerably different.

By analyzing current natural gas market, it is possible to observe that fundamental factors affecting demand or supply in the gas market have significant effects on the movements in the day-ahead gas price. Although the price of gas is still related to the price of oil, this linkage is not strong anymore (Hulshof et al., 2016).

It is also assessed that the high degree of concentration on the supply side of the gas market does not affect the gas price, suggesting that the market prices are not distorted by a lack of competition (Hulshof et al., 2016).

All the elements lead to the expectation that the period of transition of the last years is arriving at its end and a new structure for the natural gas market is delineating. In particular, the main changes are for the consumption side, because prices are now prevalently linked to the hub prices by means of supply contracts whose formulas contain a stronger reference to hub prices and a weaker link to the oil price.

The next step in the reform of natural gas market is to implement an efficient market of the transportation capacity, so that natural gas can be efficiently traded financially and physically on the hubs, which, at the moment, mainly represent a financial reference.

The optimal target would be to establish a natural gas market with the same features of the power market.

References

Bianco, V., Scarpa, F., Tagliafico, L.A. 2014. Scenario analysis of nonresidential natural gas consumption in Italy. *Applied Energy* 113:392–403.

Bianco, V., Scarpa, F., Tagliafico, L.A. 2015. Current situation and future perspectives of European natural gas sector. *Frontiers in Energy* 9:1–6.

BP. 2016. BP Statistical review of world energy. http://www.bp.com/en/global/corporate/energy-economics/statistical-review-of-world-energy.html. Accessed December 30, 2016.

Davoust, R. 2008. *Gas Price Formation, Structure & Dynamics*. Paris, France: IFRI.

Dieckhöner, C., Lochner, S., Lindenberger, D. 2013. European natural gas infrastructure: The impact of market developments on gas flows and physical market integration. *Applied Energy* 102:994–1003.

Dilaver, O., Dilaver, Z., Hunt, L.C. 2013. What drives natural gas consumption in Europe? Analysis and projection. Surrey Energy Economics Discussion Paper Series. Guildford, U.K.

EIA. 2013. Multiple factors push Western Europe to use less natural gas and more coal. Washington, DC: U.S. Energy Information Administration.

EIA. 2015. Natural gas pipelines under construction will move gas from Azerbaijan to southern Europe. Washington, DC: U.S. Energy Information Administration.

ENI. 2016. World oil and gas review 2016. https://www.eni.com/en_IT/company/fuel-cafe/world-oil-gas-review-eng.page. Accessed December 30, 2016.

EPA Ireland. 2012. Hydraulic fracturing or "Fracking": A short summary of current knowledge and potential environmental impacts. https://www.epa.ie/pubs/reports/research/sss/UniAberdeen_FrackingReport.pdf. Accessed December 30, 2016.

ETSAP. 2010. Unconventional oil & gas production. http://www.iea-etsap.org/web/E-TechDS/PDF/P02-Uncon%20oil&gas-GS-gct.pdf.

EU Parliament—DG for Internal Policies. 2011. Impacts of shale gas and shale oil extraction on the environment and on human health. https://europeecologie.eu/IMG/pdf/shale-gas-pe-464-425-final.pdf. Accessed December 30, 2016.

Goldthau, A. 2013. The politics of natural gas development in the European Union. Harvard University's Belfer Center and Rice University's Baker Institute Center for Energy Studies, Cambridge, MA.

Honorè, A. 2011. *Economic Recession and Natural Gas Demand in Europe: What Happened in 2008–2010?*. Oxford, U.K.: Oxford Institute for Energy Studies.

Hulshof, D., van der Maat, J.P., Mulder, M. 2016. Market fundamentals, competition and natural gas prices. *Energy Policy* 94:480–491.

Ikoku, C.U. 1984. *Natural Gas Production Engineering*. New York: John Wiley & Sons.

Makinen, H. 2010. The future of natural gas as the European Union's energy source—Risks and possibilities. Turku, Finland: Electronic Publications of the Pan-European Institute.

Miriello, C., Polo, M. 2015. The development of gas hubs in Europe. *Energy Policy* 84:177–190.

Paltsev, S. 2014. Scenarios for Russia's natural gas exports to 2050. *Energy Economics* 42:262–270.

Renier, P. 2013. European LNG terminal developments. *Third Annual LNG Global Congress*, London, U.K.

Richter, P.M., Holz, F. 2015. All quiet on the eastern front? Disruption scenarios of Russian natural gas supply to Europe. *Energy Policy* 80:177–189.

Stern, J. 2014. International gas pricing in Europe and Asia: A crisis of fundamentals. *Energy Policy* 64:43–48.

Stern, J.P. 2012. *The Pricing of Internationally Traded Gas*. Oxford, U.K.: The Oxford Institute for Energy Studies.

Wang, Q., Chen, X., Jha, A.N., Rogers, H. 2014. Natural gas from shale formation—The evolution, evidences and challenges of shale gas revolution in United States. *Renewable and Sustainable Energy Reviews* 30:1–28.

11

The Spanish Energy Policy Roller Coaster within the European Union: A Spotlight on Renewables

Oana M. Driha, Julián López-Milla, and Martín Sevilla-Jiménez

CONTENTS

11.1 Introduction

Energy is one of the main engines of economic development and social transformation that is present in each and every aspect of economic activity, both in production and consumption. It is therefore a basic necessity of the overall economy and a key element of the cost structure of the productive system which has a strong social and environmental impact. This is why energy policy plays a vital role in economic development, and therefore, it should take into consideration all these aspects in order to achieve its goals.

The slow development of the European Union's (EU) energy policy led to differences, divergences, and disagreements between the Member States of the EU. The absence of an effective common foreign policy and the low confidence in joint actions have resulted in an unsustainable energy policy in the long run given the challenges that the EU has to face (Marín-Quemada, 2008).

Ever since the 1950s, the EU has based its existence on energy resources and energy pacts. Nevertheless, one of the first aspects considered by the EU regarding the energy sector was its liberalization with the aim of enhancing competition within the sector. However, with the awareness of environmental changes and their consequences, the focus of the EU energy policy moved toward the environmental impact of energy generation. In other words, the EU policy started to urge the reduction of fossil fuels at the time of increasing renewable energy sources and energy efficiency. The European objective regarding renewable energy sources suggested that by 2020 each Member State should generate 20% of electricity by using renewable energy sources. Thus, each Member State of the EU triggered different mechanisms to support the development of renewables in order to reach the agreed mandatory targets. For achieving these targets, the governments had to choose the most adequate out of a wide range of strategies to suit the country's context. However, the promotion and implementation of the proper strategy is not an easy task, as many snags must be faced. One of the most common is the accumulation of deficit due to the high amount of subventions to renewable energy sources as well as the conflict between traditional and renewable energy producers and suppliers. Despite the possible barriers the support to renewables may present, recent events (e.g., the Arab spring, the accident at the oil platform in the Gulf of Mexico, the Fukushima nuclear accident), in the words of Burgos-Payán et al. (2013), underline some of the hostile consequences of using fossil fuel as an energy source: (1) supply uncertainty and price volatility of the oil and its derivatives; (2) environmental degradation and health risks; (3) huge expenditures (especially public funds) needed for mitigating the damages caused. Yet, it is worth mentioning that renewable energy has so far failed in being a protuberant competitor to fossil energy technologies as a consequence of the multiple barriers* in implementing renewable technologies (Liao et al., 2011). The high production cost and the low return on the investment of renewable energy sources compared with traditional fossil energy has contributed to the limitations of the renewable energy market (Pimentel et al., 1994). Furthermore, the use of fossil fuels may have contributed to the security of supply due to low-intensity damages on a daily basis. Having flexible fossil fuel generation in the system, the randomness of the energy from renewable sources could be alleviated. Thus, a combination of both types of energy sources is needed with a higher share of renewable sources in the energy mix.

---

* Barriers are divided mainly into four groups: (1) financial and economic, (2) institutional and political, (3) technical, and (4) awareness/information/capacity (European Environmental Agency, 2004).

The initiative of promoting renewable energy sources is not new despite the fact that the EU had issued a specific directive on this topic in 2009 (Directive 2009/28/EC). The Spanish government started its promotion more dynamically back in the 1990s by introducing an active system to jointly support renewables and cogeneration (the so-called special regime). Subsequently, a large expansion of renewables was experienced, especially of wind energy (Gelabert et al., 2011) as well as solar energy (Sebastián, 2015).

Thus, Spain climbed to a leading position among the EU Member States, following Germany, in the amount of wind power in the energy mix ranking in 2014 (ENTSO-E, 2015). This could be explained by the capacity of wind energy to generate savings that could overcome the subsidies received for promoting this technology (Azofra et al., 2014; Gil et al., 2012).

Since the end of 1996 and the beginning of 1997 until today, the regulatory process that has been affecting the energy sector has been very deep. This is not only due to the high number of directives issued, but also due to the profound changes triggered by the dynamism of the legal framework.

If during the 2000s energy policy caused many conflicts between renewables and traditional energy generators/suppliers for the enormous support of renewables, after 2020, the weaker policy framework for future renewables and energy efficiency development may not likely continue giving investors certainty regarding their investments in clean energy (Buchan et al., 2014).

11.2 State of the Art of Renewable Energy Sources

11.2.1 Global View of Renewable Energy Sources

At the beginning of the 1970s, fossil fuel was the dominant source of energy worldwide. Ever since the crisis in the 1970s, governments started to look actively for fossil-substitute energy, especially renewable energy sources (e.g., wind, solar, biomass, etc.).

More recently, fossil fuels represented more than 80% of the total primary energy and around 70% of electricity generation (IEA, 2010, 2014).* Despite this, policy makers, on one hand, and private decision makers, on the other, are searching for a sustainable transition toward a fossil-fuel-free system. This implies at least two main priorities: (1) renewable energy sources development and promotion and (2) energy efficiency. Throughout different policies and mechanisms, governments have continued to shape these aspects. This is not meant just for protecting the environment, but also for facilitating

---

* For more details, see International Energey Agency (IEA), 2014. World Energy Outlook 2014. Available at http://www.iea.org/publications/freepublications/publication/WEO2014.pdf. Accessed on October 20, 2016.

TABLE 11.1

Promoting Programs for Developing Renewable Energy

Items	Objects	Fiscal Incentive Tools	Nonfinancial Incentive Tools
Research, development, and demonstration (RD&D)	• Government • Electric producers • Grid producers	• Subsidies for research and development • Capital grants • Third-party finance	• Legislation and international treaties • Research, development, and demonstration • Guidelines for energy conservation • Public investment
Investment	• Government • Electric producers • Grid producers	• Capital grants • Bidding system • Subsidies for investment • Third-party finance • Investment tax credits • Accelerated depreciation	• Voluntary programs • Regulatory and administrative rules
Production and distribution	• Electric producers • Grid producers	• Guaranteed price • Production tax credits • Tradable certificates	• Obligations • Voluntary programs
Consumption	• Government • Consumers	• Consumer grants/rebates • Excise tax exemptions • Net metering • Fossil fuel taxes	• Obligations • Government purchases • Green pricing • Public awareness

Source: Liao, C.-H. et al., *Renew. Sustain. Energy Rev.*, 15, 787, 2011.

the access to energy for millions of people who face difficulties in this regard, as well as for creating new opportunities (REN21, 2015).

Thus, according to each country and its particular situation, different programs promoting the development of renewable energy were created and implemented (see Table 11.1).

As a consequence, a vast number of governments started planning and implementing renewable energy policies in parallel with greenhouse gas emission reduction and energy efficiency (EIA, 2007). The International Energy Agency divided renewable energy policies into nine different types as depicted in Figure 11.1.

A great part of these policies imply fiscal support from governments. These instruments have been used much earlier in some countries; for example, Denmark was already using regulatory instruments, incentives/subsidies, and RD&D for promoting renewables in 1976. This was only the beginning of the avalanche started in the mid-1990s which is actively continuing till today.

In this line, due to the undoubted awareness of the relevance of renewable energy sources and energy efficiency at the global level, a great uprising trend of policies facilitating renewable energy sources was experienced.

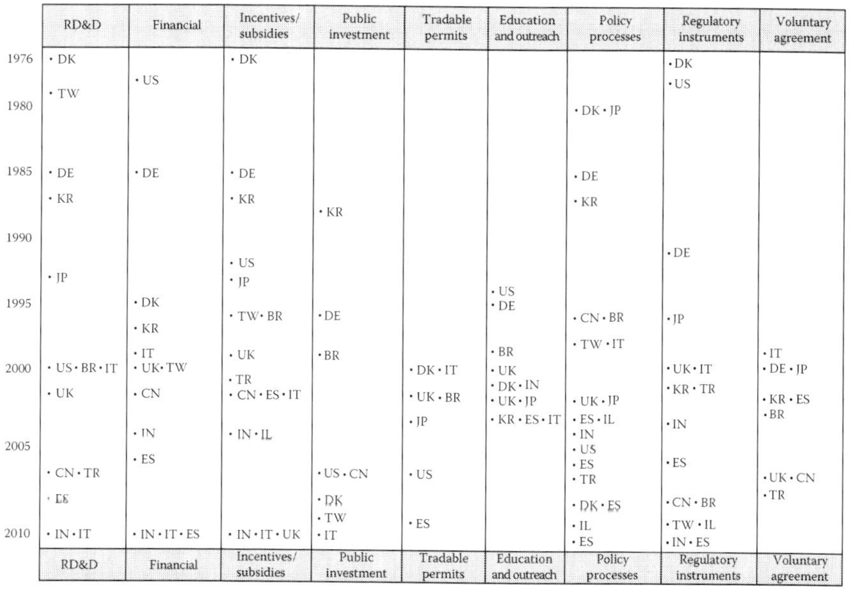

	RD&D	Financial	Incentives/ subsidies	Public investment	Tradable permits	Education and outreach	Policy processes	Regulatory instruments	Voluntary agreement
1976	• DK	• US	• DK					• DK • US	
1980	• TW						• DK · JP		
1985	• DE • KR	• DE	• DE • KR	• KR			• DE • KR		
1990	• JP		• US • JP					• DE	
1995	• DK • KR • IT	• DK	• TW · BR	• DE		• US • DE • BR	• CN · BR • TW · IT	• JP	• IT
2000	• US · BR · IT • UK	• UK · TW • CN	• UK • TR • CN · ES · IT	• BR	• DK · IT • UK · BR • JP	• UK • DK · IN • UK · JP • KR · ES · IT	• UK · JP • ES · IL	• UK · IT • KR · TR	• DE · JP • KR · ES • BR
2005	• IN • CN · TR • DE	• IN • ES	• IN · IL	• US · CN	• US	• IN • US • ES • TR	• IN • US • ES • TR	• IN • ES	• UK · CN • TR
2010	• IN · IT	• IN · IT · ES	• IN · IT · UK	• DK • TW • IT	• ES		• DK · ES • IL • ES	• CN · BR • TW · IL • IN · ES	• IN · ES
	RD&D	Financial	Incentives/ subsidies	Public investment	Tradable permits	Education and outreach	Policy processes	Regulatory instruments	Voluntary agreement

FIGURE 11.1

Mapping the adoption of various policies in representative countries. *Note:* BR, Brazil; CN, China; DE, Germany, DK, Denmark; IN, India; IT, Italy; IL, Israel; ES, Spain; KR, Korea; JP, Japan; TR, Turkey; TW, Taiwan; UK, United Kingdom; US, United States.) (From Liao, C.-H. et al., *Renew. Sustain. Energy Rev.*, 15, 787, 2011.)

In less than a decade, renewable energy policies and targets were introduced in more than 80% of the countries worldwide, registering an exponential growth. In this context, renewable energy provided the estimated 19.1% of the global final energy consumption in 2013 (REN21, 2015). Furthermore, over 58% of net addition to global power capacity was due to renewables, with China, the United States, Japan, and Germany as leaders in cost reduction as well as their significant investment in the field (around €200 billion). Wind, solar photovoltaic, and hydropower were the dominant renewable sources in 2014.

Currently, not all countries are situated in the same phase of renewable energy promotion. It is true that the majority of countries worldwide are already in the intermediary stage of developing a renewable energy market. Yet, many differences could be highlighted depending on their development stage (see Table 11.2).

A dominant instrument for promoting renewable energy sources in the EU, especially for electricity generation from renewables, is the feed-in tariff (FIT) scheme. The EU's long-term strategy is to achieve a harmonized framework for electricity from renewable energy support at EU level based on the FIT scheme (Muñoz et al., 2007). With the current legislation, EU Member States are allowed to use the support scheme considered most appropriate for each country's circumstances and socioeconomic objectives. In the Spanish case, the broad social and political coalition leading to political commitment and continuity of support schemes and the specific design elements of the support scheme itself (i.e., the FIT) are the two main factors that explain the success of its model. As a way of adapting concerns of different actors, especially the government—due to the financial impact on electricity consumers—and electricity from renewables generators, the authorities decided to modify the regulatory framework regarding FIT several times.

However, it must be noted that, generally, the support of electricity from renewables is finally paid by electricity consumers as part of their electricity bills (Sáenz de Miera et al., 2008) despite the fact that Jensen and Skytte (2003) underlined that higher electricity from renewables deployment would incur a reduction of final electricity prices. This is because the promotion of electricity from renewables encourages its generation because of lower variable costs than fossil fuel conventional electricity. Wholesale electricity price is generally established by fossil-fuel-fired plants, which are usually the marginal generation plants. At the same time, these types of plants are substitutes for renewable energy sources. Therefore, the wholesale electricity price would be reduced with a higher deployment of renewables. This reduction could balance the growth in final electricity prices as a consequence of renewables support, leading to a net reduction in retail prices. In other words, renewables promotion could lead to a win–win situation while an increase in renewables deployment (counting its environmental and socioeconomic benefits as well) could contribute to a reduction in electricity prices (Sáenz de Miera et al., 2008).

TABLE 11.2

Goals and Energy Instruments Characteristic to Each Renewable Energy Market Stage of Development

Phase	First Stage: Undeveloped Market	Second Stage: Developing Market		Third Stage: Developed Market
Steps	*R&D, investment*	*Production*	*Consumption*	*Production, consumption*
Goals	• To establish renewable energy market	• To improve the production of renewable energy	• To improve the consumption of renewable energy	• To replace fossil fuel with renewable energy • To return to the free market mechanism
Non-market-based policies	• Regulatory instruments • Policy processes • Voluntary agreement • Education and outreach			
Market-based policies	• RD&D	• Financial • Incentives/subsidies • Tradable permits	• Financial • Public investment	• Liberalization
Specific applications	• R&D grants and subsidies • Demonstration	• Investment deduction • Tax credit • Accelerated depreciation • Guaranteed price • Obligations and tradable permits	• Tax incentives • Grants and subsidies • Public investment	• Removal of fossil energy subsidies • Carbon tax • Green pricing • Removal of renewable energy incentives/subsidies
Mechanisms	Quota system	Quota system Price system	Price system	Free market system

Source: Liao, C.-H. et al., *Renew. Sustain. Energy Rev.*, 15, 787, 2011.

11.2.2 Relevance of Renewable Energy Sources in the EU

The EU triggers the implementation of an eclectic range of policies focused on climate change mitigation and security of energy supply. These policies are mainly centered on emission-trading schemes, support schemes for electricity produced with renewable energy sources, and measures to encourage energy efficiency. While the emission-trading schemes have been functioning since 2005, each Member State handles electricity from renewables and energy efficiency strategies. According to del Río González (2010), overlaps, conflicts, and synergies in the interaction and combination of these instruments raise serious concerns.

The European electricity market was fully liberalized on July 1, 2007, with the aim of raising competition and decreasing electricity prices for consumers (including households) even though the liberalization process had started in the 1990s. Electricity market reforms began by opening markets in wholesale markets, followed by generation and transmission, and then going toward retail supply to household consumers (Moreno et al., 2012). Although the aim of these reforms was the liberalization of generation and retailing activities and, as a consequence, the decrease in electricity prices, it seems that the main objective, price decrease, was not achieved as expected.

The liberalization of the European generation market should have led to changes in the market structure involving a clear reduction of market concentration in many Member States of the EU. Yet, entry barriers intensify the difficulty of generating monopolies. Generally, well-established companies have the advantage of transport networks from their locations, which has high strategic value. Additionally, grid connection acts as the most problematic entrance barrier to developing electricity from renewable energy sources. This is even more crucial if the integration of plants for electricity generation from renewable energy sources entails grid reinforcement. An important decrease in the electricity production market share of the largest generators was experienced especially in Spain and Italy during the last 15 years (see Figure 11.2).

Since 1996, with the publication of the Green Paper on renewable energy sources, the EU started to outline more explicitly its position in this context (Jones, 2010). One year later, after the White Paper of renewable energy sources (European Commission, 1997), the EU tried to establish a European framework for energy (Sevilla et al., 2013). Consequently, the Directive 2001/77/EC* set challenging indicatives for national goals to almost double the share of energy from renewables in the EU electricity mix from 12% in 1997 to 21% in 2010. With the recent global economic crisis, and hence the need of revising the EU's growth strategy (i.e., EUROPE 2020), the increase of the share of renewable energy sources at EU level was proposed in terms of energy consumption up to 20% by 2020 (for more detail see European Union (2009)). Only 8 years

---

* For more details, see Directive 96/92/EC and Directive 2003/54/EC (European Commission, 2003).

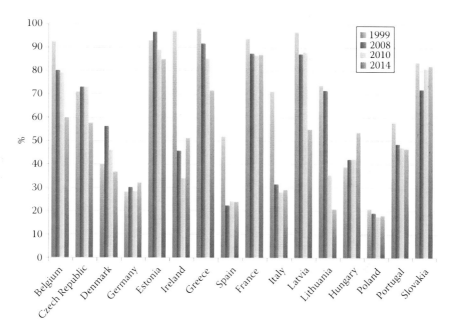

FIGURE 11.2
Market share of the largest generators in the electricity market (in %). (From Eurostat, http://ec.europa.eu/eurostat/data/database.)

later, a European directive was developed and approved, establishing national goals as well as the path to follow in order to achieve 20% electricity generation from renewable sources by 2020 at EU level. Many differences emerged within the EU (Menanteau et al., 2003) as the plans for supporting renewables and promoting network connections are designed using National Action Plans.

It must be underlined that the EU is one of the leaders in promoting renewable energy sources and in explaining to society how the consequences of climate change will affect the environment both from an economic and a social focus. In line with the EU green objectives,* one of the main targets the EU has to face is to raise the share of renewable energy for electricity generation. Despite the fact that the politicians who set the EUROPE 2020 growth strategy had a well-defined objective of tackling climate change, the 2020 renewable targets, at EU and national levels, have been much criticized for producing expensive forms of carbon drop.

In October 2014, the EU leaders established even more ambitious targets by 2030: 40% reductions in greenhouse gas emissions (compared to 1990 levels),

---

* According to the green objective of the EU growth strategy (EUROPE 2020), by 2020 the EU should experience a fall in greenhouse gas emissions by 20% compared with the levels of 1990, should have 20% of energy from renewable sources, and should increase the energy efficiency by about 20% (not binding).

27% renewable share of overall energy consumption, and 27% improvement in energy efficiency compared to business-as-usual energy projection.* These issues notwithstanding, the new targets set for 2030 do not seem to imply a great extra effort (Buchan et al., 2014). Continuing the present policies for another decade, the results would consist of 32% greenhouse gas emissions reduction and 24% renewable share of energy, while energy efficiency would barely improve.[†] More precisely, Bulgaria and Estonia have already achieved the EU objectives[‡] in 2012, while Sweden overpassed it (Figure 11.3).

There is a wide range of strategies applied according to each country's characteristics and political preferences. And yet, which instrument is the most appropriate for increasing the dissemination of electricity from renewables

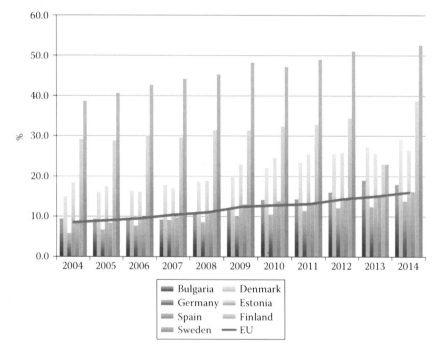

FIGURE 11.3
Share of renewable energy sources in the gross final consumption. (From Eurostat, http://ec.europa.eu/eurostat/data/database.)

---

* http://www.consilium.europa.eu/uedocs/cms_data/docs/pressdata/en/ec/145397.pdf.
† http://eur-lex.europa.eu/legal-content/EN/TXT/PDF/?uri=CELEX:52014SC0016&from=EN.
‡ Even if the EUROPE 2020 objective in the field of renewable energy is to reach 20% share of renewable energy sources in the final consumption, this does not mean that each Member State must have 20% of renewables in their final consumption. Each country has a different objective, which, by 2020, will allow the EU to have, on average, 20% of energy from renewables. For more details, see http://ec.europa.eu/europe2020/pdf/targets_en.pdf.

is still a topic of controversy. The question remains in deciding between feed-in tariffs and tradable green certificates based on quotas (Haas et al., 2011).

Under the umbrella of environmental and socioeconomic reasons, public support for electricity from renewables has been justified. Negative externalities have been diminished due to much lower pollution through electricity from renewables in comparison with traditional electricity from fossil fuels; hence the objectives of the Kyoto Protocol* in terms of climate change mitigations are being pursued and the emissions of local pollution have been moderated.

By 2050, an almost carbon-free EU power must be in place (IPCC, 2007). Renewables have to play a vital role in achieving this goal as energy efficiency is considered crucial in reducing or maintaining power consumption and greenhouse emissions, even if it is not capable of fulfilling the green objectives alone (Jacobsson et al., 2009).

After the shock provoked by the world crisis of 2008, energy consumption suffered a high decrease. Nevertheless, due to steps undertaken by governments in designing mechanisms for recovering from this crisis, energy consumption recouped considerably. In 2014, renewables continued their development despite the drastic decrease in oil prices.

Additionally, while renewables installed capacity and energy production extended considerably, investments in renewables were higher than investments in plants of fossil fuels. Electricity was the subsector with the fastest and highest increase of renewable capacity, with wind, solar photovoltaic, and hydropower being the dominant technologies.

11.3 Spanish Focus on Renewables

The tremendous dependence on energy imports, as a consequence of the low domestic energy production, and the environmental issues, pushed the EU, especially Spain, to develop policies to promote the use of energy from renewable sources and achieve great energy efficiency without losing control of energy consumption. Thus, many changes were established not only at EU level (Ruiz Romero et al., 2012) but also at the national level (del Río González, 2008) regarding the production of electric power using renewable energy. In this regard, Spain was highlighted as one of the most successful countries in the public promotion of electricity from renewable energy sources, especially from wind (del Río González, 2008). The promotion strategy of the Spanish authorities was even acclaimed by the European Commission due to its effectiveness and lower costs in comparison with other countries (del Río González, 2008; European Commission, 2005; Held et al., 2006). The leader of

---

* United Nations Framework Convention on Climate Change (UNFCCC, Kyoto, December 11, 1997).

the EU in electricity from renewables is Spain,* followed by Germany, which together made possible the increase of overall electricity from renewables capacity of the EU.

Spain is known as a pioneer at the European level in promoting renewable energy sources as well as a leader in this field. This is because Spain became the country with the highest support granted to renewables, surpassing even Germany not only in the degree of penetration of these energy sources in the energy mix—about 30%—but also on the average cost of such support (Sallé-Alonso, 2012). The leading renewables in generating electricity are mainly wind and solar photovoltaic technologies, while biomass and hydro are deadlocked (del Río González, 2008). While there is good progress in the renewable electricity target, when it comes to biofuels, Spain is rather far from the expected target (Solorio, 2011). Therefore, the Spanish recipe for achieving the European objectives is not entirely satisfactory. Furthermore, the strategy of promoting renewables in Spain led to several issues such as huge accumulation of tariff deficit as well as a fall in capacity payment in many legal issues, and, lately, as a consequence, there is a controversial arbitration process† between investors in renewable energy and the government due to cuts in the support of renewables.

11.4 Stages of the Renewable Energy Policy

Spain, as well as the whole world, but especially the EU, is experiencing a clear trend in increasing the share of renewable energy since the 1990s. The case of Spain can be considered to be, up to a certain point, unique due to its huge list of energy laws and the large imbalances created by these. Three different stages can be underlined in the Spanish renewable energy policy as depicted in the following sections.

11.4.1 Initiating the Support of Renewables

Unlike the EU, Spain was suffering the consequences of a huge dependence on fossil fuels. This situation continued over time as the dependence of Spain on energy imports from the 2000s until now is over 70% while the EU average is around 50%. Therefore, the need to reduce this dependence pushed the Spanish government to make strategic decisions.

---

* Among others, Dinica (2003), Bustos (2004), García and Menéndez (2006), Bechberger (2006), and Meyer (2007) underline the effectiveness of the Spanish feed-in tariff model.
† http://www.tribunalconstitucional.es/es/salaPrensa/Documents/NP_2016_001/2013-05347STC.pdf.

As a consequence of the geopolitical changes in the Middle East in 1973 and of the second crisis of 1979, oil prices increased exponentially as well as the insecurity of supply, among others (Folgado-Blanco, 2003). In Spain, apart from the price climb, a notable expansion of demand was experienced. All these were pinpointing toward a growing fear of a supply crisis. Because of its high dependence on energy imports, ever since the beginning of the 1980s, the Spanish government had to develop a strategy for promoting alternative energy sources. At first, national carbon, hydropower, and nuclear power were supposed to make up for the reduction of oil use in Spain. Nevertheless, this was not enough. Thus, Law 82/1980 for the Conservation of Energy initiated the promotion and support of other alternative sources of energy as renewables.

At the beginning, through the Royal Decree 2366/1994, electricity produced by hydro sources, cogeneration, and renewable energy sources was regulated. According to Sevilla et al. (2013), this was the beginning of the establishment of the basic contractual relationship between electricity from renewables producers and distribution companies. Thus, distributors have to buy the electricity surplus of plants with an installed capacity of less than 100 MW.

11.4.2 Development of Specific Technologies and Strong Support of Renewables

One of the main objectives of the EU and its Member States was the liberalization of the energy market in order to ensure higher competitiveness and, hence, a better and sustainable economic development. Thus, the liberalization of the electricity market was postulated through the European Directive 96/92/EC. The same directive had to be extended in the Spanish context by a new law of the electricity sector (Law 54/97). This was the beginning of the renewable energy sources support in Spain. A triple objective can be detected in this law: (1) ensuring power supply, (2) ensuring the quality of the supply, and (3) ensuring the lowest cost possible. Additionally, new aspects were established by means of this law as producers of electricity from renewable sources started to enjoy the following:

1. A different treatment under the "special regime"—further developed—aimed at fulfilling the EU target of gross energy consumption (12% by 2010 and 20% by 2020).
2. Guaranteed grid access.
3. Price support. For plants less than 10 MW a premium was set and financed by the government. The support level will be determined based on the effective contribution of electricity from renewables to greenhouse gas emissions, primary energy savings, energy efficiency, and investment costs.

The development and effective implementation of a new regulatory framework that set the beginning of the renewables era in Spain was achieved not only with the entrance into force of Law 54/1997 of the Electricity Sector but also with Law 34/1998 on the hydrocarbon sector. According to Folgado-Blanco (2003), both laws defend the basic elements of the Spanish energy policy as they

- Promote economic growth, so that the power supply will not be an obstacle in the real convergence with the hard core of the EU (most prosperous Member States).
- Ensure security of supply with an appropriate quality and at an affordable price to the entire population, despite the very high external dependence.
- Harmonize the use of energy with effective protection of the environment, so that sustainable long-term development can be reached.

These new trends established by the legislation of the late 1990s ensured the prevailing principles of freedom of installation and contracting. At the same time, the new framework showed more transparency and allowed higher competitiveness, undoubtedly due to the enhancement of the energy markets. In this legal context, the Spanish government was looking for an appropriate balance between sustainable development and energy policy measures. In other words, the aim was to design an energy model tailored to the needs and characteristics of the Spanish market to avoid possible "bottlenecks."

Moreover, the avalanche of changes continued with the Royal Decree on Special Regime (RD 2818/1998). Two different options were offered to electricity generators from renewables: fixed-premium on top of the electricity market price and fixed-feed-in adjusted annually. The fixed-feed-in system gave investors the opportunity to know their own revenues in advance, independently of the market price shifts. As a consequence, huge imbalances were caused as well as an overload of the final price for consumers (del Río González and Gual, 2007). According to Robinson (2015), the overload of the final price was due to an improper strategy of promoting renewable energy sources and, hence, to the tremendous governmental wedge.

Furthermore, a Plan for Promoting Renewable Energy Sources IDAE (1999) was established in 1999 that set more explicit objectives. The aim of this plan was to maintain the commitment to meet at least 12% of the total energy use from renewable sources as well as achieving 29.4% of electricity generated from renewable sources and 5.75% of transport fuel needs to be met from biofuels by 2010.

In the field of biomass and wind electricity, the targets were established through the Electricity and Gas Infrastructure Plan for 2002–2011. Very ambitious targets were established for each administrative region.

The framework of the Royal Decree 436/2004 allowed generators of electricity from renewables to sell it to distributors at a fixed tariff or directly to the market at the market price plus a bonus or with even better conditions. In both cases, a support based on the average electricity tariff was included, which was annually set by the government. This was another incentive for encouraging the participation of electricity from renewables in the wholesale electricity market. The aim of this RD was to increase electricity from renewables generators selling their electricity directly to the market, causing a high increase in governmental expenditures and imbalances.

An even higher target for electricity from renewables was established in 2005 through the Renewable Energy Plan for 2005–2010 IDAE (2010), reaching 30.3% by 2010. Targets for thermal applications and biofuels were also defined, forecasting an investment of €1.156 million between 2005 and 2010 and a public support in the same period of €2.855 million.

If previous legal frameworks linked electricity from renewables to the average electricity tariff, the Royal Decree 661/2007 proposed, instead, their disengagement considering the evolution of the Consumer Prices Index for updating the support. Consequently, for renewables support, a cap-and-floor system was implemented (del Río González, 2008) without changing the previous supporting system of renewables established by RD 436/2004. Nevertheless, with this new focus, an increase in the economic incentives for this kind of investment was created, emphasizing the expectations of renewable technologies (Agosti and Padilla, 2010). Solar photovoltaic became one of the most attractive renewable technologies and this RD facilitated the installation of more than five times the target for 2010 only 1 year after its approval. This was classified as a remarkable promotion of solar thermal industrial activity, which fixed a 0.27 h/kWh feed-in tariff for the electricity generated by solar thermal technologies. Combined with the possibility of constructing mixed plants with gas, this generated a great interest for solar concentration technologies among investors and the Spanish industrial sector (Caldés et al., 2009). This strategy was the outcome of the belief that solar energy is one of the most promising sources of clean energy, especially in countries like Spain where solar radiation and solar electricity generation potential is remarkable.

Despite the supposition of Jensen and Skytte (2003) that an increase in the renewables deployment could lead to a decrease in prices, a continuous increase of electricity prices was experienced in both domestic and industrial sectors. Currently, the legal aspect of the development stage of renewable energy support is considered by the former Industry Minister, Miguel Sebastián, as the beginning of a set of energy errors, especially in the solar field (both photovoltaic and thermal) as a consequence of energy excess (Sebastián, 2015).

Nonetheless, due to a rather ambitious regulation framework that caused huge imbalances, the government was obliged to make considerable changes to regain the support of renewable energy sources.

11.4.2.1 Slowing the Support of Renewables

The need of mitigating the enormous imbalances created through the legal framework of renewables without a realistic strategy in the long run to meet the characteristics and needs of the Spanish market led to a third stage: a slowing down in the support for renewable energy sources. The government had to take its foot off the pedal as the incentives for investing in renewables led to a too large public deficit.

The first step back in supporting renewables was undertaken with the Royal Decree 1578/2008, which was meant to rationalize the support of electricity from renewables from PV by moderating it. The appearance of disincentives for investing in renewable energy sources created clear problems between investors, producers, and the government. Despite this cut in the incentives of supporting renewable energy sources, the Spanish system was a step ahead as it has seen great achievements in participation in the national electricity mix (42% of the total generation capacity was registered in 2008 and 22% of total electricity production was coming from renewables). Unfortunately, it was not efficient enough because of the relatively high costs of production. The support scheme was not sustainable and a new reform of the incentives was needed.

Another reduction in the incentives of renewables support was carried out through the Royal Decree Law 1/2012 by suppressing the preassignment procedures of incentives for newly installed plants of electricity generation from cogeneration, renewables, and waste. In this context, Collado (2012) underlines that the internal rate of return for PV plants installed was 6.75% in 2011 with a coverage ratio of debt service close to the technical default. Not all types of technologies are liable to these considerations. The solar photovoltaic technology, with the initial incentives, caused major imbalances and created a bubble based on too generous a public subsidy. Under a brutal economic crisis, where the Spanish legal international credibility was questioned, additional problems were added due to energy planning and the implementation of commitments set in the European Directive 2009/28/EC on the promotion of the use of energy from renewables. The details were included and explained in the Renewable Energy Plan 2011–2020. Once more, given the complexity and the extent of the situation resulting from the incentives to renewables, further steps were required for getting to a balanced position.

The Royal Decree-Law 2/2013 implemented urgent measures in the electricity and financial sectors toward special regime installations (renewables and cogeneration) in order to correct the tariff deficit and reinforce financial stability (CNMC, 2014). Hence, this new regulation eliminated the reference premium for all technologies, amending in this regard the RD 661/2007. Therefore, the options were now the following:

1. A regulated feed-in tariff
2. Selling electricity at market price with no additional premium

Additionally, the inflation index used until then for updating the remuneration of regulated activities started to eliminate variations in energy and food products and any impact as a consequence of tax changes.

The reform of the electricity in Spain started with the Royal Decree Law 9/2013 targeting the promotion of efficiency, transparency, and competition and the reduction of the tariff deficit and the financial stability in the electricity market (CNMC, 2014).

Law 24/2013 of the Power Sector obliges all renewables installations to sell the produced electricity in the market, receiving the market price together with regulated revenue. Thus, the parameters to determine regulated income, according to this act, have to be reviewed every 6 years.

Likewise, Law 24/2013 establishes three different categories of self-consumption. Additionally, grid connection and extension costs should be considered while integrating electricity from renewables generation technologies into an existing network (Swider et al., 2008). This aspect is also covered by Law 24/2013 as it obliges those installations connected to the grid to contribute to the costs and services of the system in the same conditions as the rest of the customers.

The Royal Decree 900/2015 regulates the administrative, technical, and economic conditions and generation for self-consumption, keeping in mind the relevance and need of the grid system in which self-consumption must be regulated. Thus, small self-consumers are forced to give away the energy that they do not consume to power companies. Furthermore, self-consumers connected to the grid are already paying the entire fixed portion (having to pay the same system tolls as any other consumer) and the portion corresponding to the energy that they may demand from the grid. This could be interpreted as a barrier to the self-consumption development in Spain.

The Royal Decree 947/2015 was approved for supporting new installations of plants for generating electricity from biomass and wind energy.

Although steps were taken to regain support for renewable energy sources, the price of electricity continued to ascend. This increase was not only the result of the rising energy and supply price but was caused especially by the growing costs of taxes and levies as well as of networks (see Figure 11.4). It is therefore not surprising that more than 50% of the price of electricity is composed of network costs together with taxes and levies. This seems to be the result of the policy undertaken in the second stage of renewables support (end of the 1990s until 2007), among others.

In spite of the increase in network costs and taxes and levies for industrial consumers (see Figure 11.5), it must be underlined that their proportion in the electricity price is much lower than that of domestic consumers.

Consequently, and contradictory to the supposition of Jensen and Skytte (2003), renewables support seems to lead, at least in the Spanish scenario, to an increase in final electricity price. Is this because Sáenz de Miera et al. (2008) had not yet accomplished their study? Further research should be

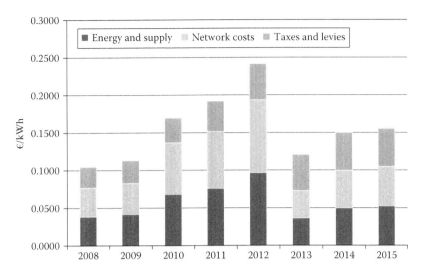

FIGURE 11.4
Evolution of electricity price components of domestic consumers. Band Dc: 2500 kWh < consumption < 5000 kWh. *Note*: Band Dc is chosen as a medium domestic consumer. (From Eurostat, http://ec.europa.eu/eurostat/data/database.)

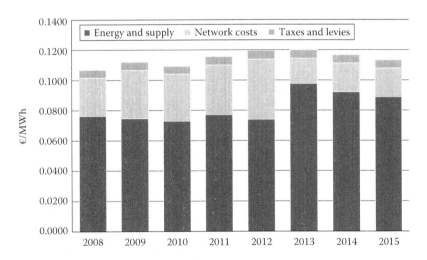

FIGURE 11.5
Evolution of electricity price components for industrial consumers. Band IC: 500 MWh < consumption < 2000 MWh. *Note:* Band IC is considered as a medium-size industrial consumer. (From Eurostat database, http://ec.europa.eu/eurostat/data/database.)

carried out in order to give a clear and tested answer to this question. But meanwhile, the huge imbalances caused by the inadequate use of the feed-in systems has led to an increase in electricity prices, which is ultimately suffered by the consumers, as del Río González and Gual (2007) state. This situation is even worse when it comes to domestic consumers, as Robinson (2015) underlines.

Yet, the results of the drawback stage in the excessive support of renewable energy sources have started contributing to the decrease in electricity prices since 2013.

11.4.3 Counterpart of Renewables Support

As a consequence of the strong support of renewable energy sources by the Spanish government since the 1990s, two main issues can be underlined: (1) tariff deficit and (2) security supply. These two aspects can explain the need for the fulminant changes experienced in the policies for renewable energy sources and their support.

11.4.3.1 Tariff Deficit

Promoting and supporting renewable energy sources meant an increase in tariff deficit for the Spanish government.* Since the early 2000s, especially since 2005, a growing trend has marked the beginning of a new stage, which resulted in a tariff deficit (cumulative) of more than €32,000 million by the end of 2012. Despite the sharp increase in the rate paid by consumers in recent years and the efforts to find the optimal recipe in energy regulation, the increasing importance of renewable energy sources in the energy mix has contributed considerably to the increase in tariff deficit and is expected to follow the same path in the future (Fabra and Fabra, 2012).

Sallé-Alonso (2012) pinpoints that the tariff deficit could have been solved with small adjustments with a frequency (of 2, 3, or 6 months) adapted to the size of the imbalance detected. In his opinion, the government has four different regulative keys as depicted in Figure 11.6. Accordingly, an improper management of the four keys is the reason for tariff deficit accumulation in the Spanish system.

With the increasing tariff deficit, and despite the established principle of tariff sufficiency in Law 54/1997 to be accomplished by the public administration, new legal frameworks have been developed. Nevertheless, with the shifts experienced by the energy legislation focused on the reduction of the

---

* Tariff deficit is the difference between the recognized rights of incomes and the electricity tariffs.

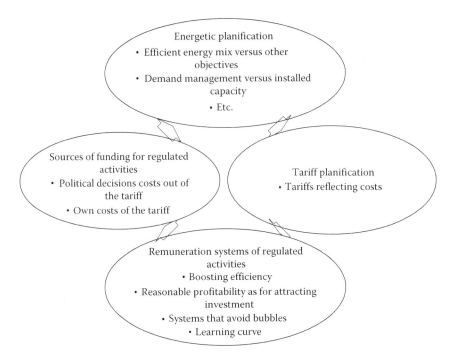

FIGURE 11.6
Regulative keys of public administration. (From Sallé-Alonso, C., *Papeles de Economía Española*, 134, 101, 2012.)

tariff deficit through numerous legal changes since 2008; the trend changed completely in 2014 as a surplus was experienced (see Figure 11.7).

Even though the strong support of renewables came to an end with the RD 1578/2008, a long list of laws, royal decrees, and royal decree laws had to be designed and approved in order to solve the consequences of the lack of a realistic energy strategy with a long-term focus on renewable energy sources.

As a consequence of the lack of consistency in the administrative actions with the RDL 6/2009, the Royal Decree Law 6/2010 defined further increases in deficit limits that had previously established the Royal Decree Law 6/2009.

The Royal Decree Law 1/2012 tried to stop future high costs planning that would result from expensive renewable technologies. Alongside this, several other royal decrees were approved in the same year with a similar aim: to reduce tariff deficit (RDL13/2012 or RDL 20/2012).

In addition, Law 24/2013 of Power Sector sets forth the legal background of this sector keeping in mind the crucial need to avoid the accumulation of new tariff deficits.

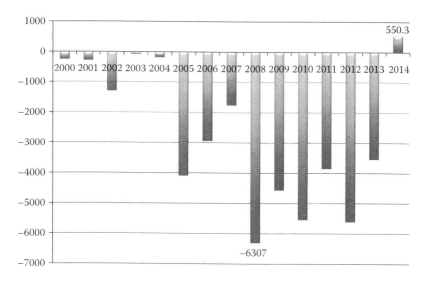

FIGURE 11.7
Evolution of the tariff deficit in Spain (2000–2014). (From Comisión Nacional de los Mercados y la Competencia (CNMC), Spanish Energy Regulator's National Report to the European Commission 2014, https://www.cnmc.es/Portals/0/Ficheros/Energia/Publicaciones_Anuales/140731_NR_CNMC_final.pdf, 2014. Accessed November 20, 2016.)

11.4.3.2 Security Supply

The downturn of electricity demand suffered all over the world was even stronger in Spain due to the extreme economic crisis it faced. In such a context, with a lower demand today than in 2006, ensuring power supply was not an easy task. The strong support of renewable energy sources promoted since the 1980s until late 2000s (RD 661/2007) contributed to the profitability issues faced by many power plants. This background facilitated a contradictory perspective of traditional and renewable energy agents.

In the energy field, strong investments were made since 2003 mainly in renewable energy sources, especially wind and solar photovoltaic, as well as in power plants, this time essentially in combined cycle gas turbine (CCGT). In spite of this, the limitation of the installed capacity came as a consequence of the global economic crisis pictured in the 25,348 MW of CCGT at the end of 2015 and 22,845 MW in wind energy (REE, 2015). The need to increase the penetration of renewables and of coal in the electricity mix was combined with a high reduction in the functioning hours of CCGTs to 10% in 2015. Additionally, a clear installed-capacity surplus was observed in CCGTs. Hence, several changes were requested in the legal framework regarding capacity payment.

Guaranteeing energy supply in order to avoid possible blackouts was effected by encouraging investments in new power plants as well as ensuring the availability of the necessary plants for guaranteed electricity supply (plants with large and stable capacity). For this purpose, the capacity payment mechanism was created under Law 54/1997. The design of the Spanish electricity market had to show a clear support for electric energy producers through a regular additional income (different from the one obtained due to energy trade) in order to help them recover fixed costs. Both availability and installation should be considered in the medium and long terms for calculating capacity payment.

The capacity payment mechanism was maintained by Law 17/2007 and was established according to the capacity required by the system. Moreover, the Ministerial Order ITC/2794/2007 developed the concept of capacity payment taking into consideration that the energy demand is inelastic and that the grid is not perfect. This mechanism was structured in two parts:

1. The investment incentive is an incentive in the long run designed for promoting the construction and the effective use of new installations for generating electricity through payments that will allow the promoters to recover their investment. It is linked to the first 10 years of its use or to ecological investments during the first 10 years subject to a minimal annual availability (over 90% in tariff period 1). Hence, ordinal regime peninsula power plants (located in the mainland, not in the islands) with an installed power of at least 50 MW had the right to an annual payment of €20,000/MW/year for 10 years since start-up. It also envisaged the possibility of auctions for the allocation of investment incentive if the coverage ratio (CR) was less than 1.1, but the scheme based on the CR was not implemented.

2. The availability service is a service for the medium run for completing the services for the system's adjustment used for guaranteeing the supply in the short run. It is meant to offer to the system operator the specific capacity for a period up to 1 year of those generating power plants, which might be unavailable during periods of highest demand. This could be because the energy market does not allow them to recover their fix costs under their regular functioning (e.g., fuel thermal power) or because the raw material used can be stocked at a low cost or because there are technologies in which the raw material can be stored at low cost. A certain level of uncertainty regarding the specific collection volume distribution and its timing (adjustable hydraulic installations) exists. This payment was not fully developed in this regulation, but it was temporary applied in the period of January–July 2008. The economic endowment was up to €137 million.

The important drop in the hours of functioning of CCGTs, the decrease in electricity demand, the significant investments in CCGT and renewable energy resources (RES) plants, a high coverage rate, and the installed capacity surplus are some of the incentives to modify the legal framework regarding the availability of service payments. In the Royal Decree 134/2010, restrictions set for capacity payment were approved. Since then, power plants employing domestic coal were programmed by REE in order to meet the annual targets. The decision to promote the use of domestic coal was because electricity generation with this fuel decreased significantly since 2008 due to reduced competitiveness and sought to maintain the operation of those electricity production units that employed local primary energy sources, adopting measures to prevent alteration of the market price. These plants acquired guaranteed minimum annual hours of operation, which resulted in the reduction of electricity production by CCGTs, with a drop in demand and an increase in RES production. The main consequences of this regulation are the increase in greenhouse gas emissions and the lower production of other technologies installed, which may jeopardize its continuity.

For promoting the availability of plants under the ordinal regime for a period of 1 year, the Ministerial Order ITC/3127/2011 developed and established the availability service of €5150/MW according to the technology (CCGT having the highest incentive). Additionally, the investment incentive was raised to €26,000/MW/year for a period of 10 years. Hence, the availability service is the availability to the system operator of the capacity of the whole production network or of a part of it. More precisely, it is the availability of the capacity of those thermal installations that produce electric energy under the ordinary regime described in the First Section of the Installation of Production of Electric Energy Administrative Register, which might not be available at the moment of maximum demand without this incentive. This might be because it is a marginal technology of the daily market, that is, of fuel-oil plants, combined cycle, and coal plants, and also of pure water systems that have mixed pumping and reservoir (damming). Availability service as developed in this regulation is a transitory instrument applicable between December 2011 and December 2012. Additionally, the investment incentive for the long run for plants whose start-up certificate was issued after January 1, 1998, was revised with the aim of updating and adapting this kind of payment to the changes that occur during the operating hours of these centrals, which makes this payment maladjusted and includes in this service centrals with significant environmental investments for reducing emissions of sulfur oxides (SO_2), in addition to the desulfurization plants that were already considered in the 2007 legislation.

The Royal Decree-Law 13/2012 modified the values of investment incentives to €23,400/MW/year for 2012 alone, justifying this modification by the existence of a low demand of electricity and a low risk of installed capacity deficit.

The Royal Decree-Law 9/2013 established an indefinite reduction of investment incentives with a value of €10,000/MW/year; this applies to all new production facilities, except those that obtain a certificate of final commissioning before January 1, 2016, in which case they are entitled to €10,000/MW/year for 20 years. Facilities that existed before this regulation came into force, and that are entitled to this incentive, received the incentive for double the term in order to cover the 10-year period for which they were entitled to receive an incentive according to the regulation of 2007. In 2013, a proposed RD was designed for capacity mechanism and hibernation* by amending certain aspects of the electricity market to the regulated changes that took place in 2013. With regard to capacity payments, the current mechanism is reviewed, stating that the financing capacity payments correspond to suppliers and direct consumers in the market. Regarding the investment incentive for the long term, it has been reviewed and an auction mechanism has been established for the new facilities to be implemented in the event that the system operator detects a shortfall in covering demand in the long run, depending on the coverage ratio calculation. The procedure and the requirements for allocating investment incentives have been detailed. In the case that the CR for the next 4 years is below the established minimum, a call auction will commence. Transitional arrangements have been made for facilities that existed prior to this RD to receive the investment incentive of capacity payments.

In October 2015, the National Energy Authority (Comisión Nacional de los Mercados y la Competencia) gave negative information regarding the capacity payment of what pretended to be a subvention for the modernization of thermal centrals, which was planned as an oxygen bubble requested by the mining sector guaranteeing the sale of national carbon to thermo centrals, which in turn would receive subventions for investments in adapting their installations to the new European legislation for industrial emissions.

In the last modification of the energy regulation, the Ministerial Order IET/2735/2015 establishes the fees for electric energy access for the year 2016 and approves different types of installations and parameters for incentives for electricity production installations from renewable sources, cogeneration, and waste. This regulation establishes a unitary price for financing the capacity payment applicable to electricity consumers launched in 2007 and revised in 2011 under a special regime. Under reasonable parameters (regarding the number of agents), the level of capacity resulting from

---

* During hibernation, a temporal shutdown of the plant is carried out. This makes it possible to auto-adjust the latent capacity excess due to suppliers' decisions; all this without damaging either the system or its supply safety. The competitive auction procedure regulates the assignment of capacity susceptible to hibernation. Normally, the period of temporary hibernation of CCGT plants is 1 year. Therefore, at least 6 months beforehand, an auction for each period is organized under the supervision of the CNMC. Regarding the liquidation of the auctions, Red Eléctrica Española is the authority responsible for this process, while the administrative procedures run under the State Secretariat of Energy.

private decisions falls far short of the social optimum (Castro-Rodríguez et al., 2009). Two regulatory mechanisms—capacity payment and price-adder—that used to generate supplementary incentives for private agents to install capacity are considered to be ineffective and/or unduly expensive. The CP method consists in awarding each generating unit a daily payment—when it is available—which is computed by multiplying the firm capacity of each generating unit by the per unit capacity payment (€/MW), which may be uniform or may vary with the season (Batlle et al., 2007). In this context, two main weaknesses of this mechanism are highlighted by Batlle and Pérez-Arriaga (2007): (1) the low capacity of providing generators with an incentive to make a special effort to be available and to produce electricity in situations of real need and (2) the lack of a guarantee of a reasonable volume of installed capacity to satisfy the demand every time.

Up until today, there is no consensus in the literature regarding an adequate model for calculating the actual firm capacity of the different power-generating plants.

Summing up, the regulation focused on capacity payment is depicted in Figure 11.8.

With the aim of guaranteeing financial stability in the electricity system as an essential requirement for ensuring its economic sustainability and security of supply, the Royal Decree Law 9/2013 set out a series of broad measures targeting the following aspects:

- To establish a regulatory framework that will guarantee financial stability in the electricity system
- To remove deficit from the electricity sector, prevent future deficit, and guarantee supply to consumers at the lowest possible cost and with increased transparency
- To simplify and clarify electricity bills and encourage competition in domestic electricity tariffs, while maintaining the discount known as the "social bonus" for vulnerable customers

Moreover, Law 24/2013 of Power Sector was aiming to guarantee electricity supply at an adequate level of quality and at the lowest price possible. The economic and financial sustainability of the system and the effective competition, together with environmental sustainability, continue to be the spotlight of the legal framework.

Furthermore, the investigation of the Capacity of payment (CP) mechanism in 2015 as an illegal subvention of the energy system by the European Commission must be underlined. CP is considered to be a mechanism for financing power generators only for the security of supply, but not as a solution to their low profitability.

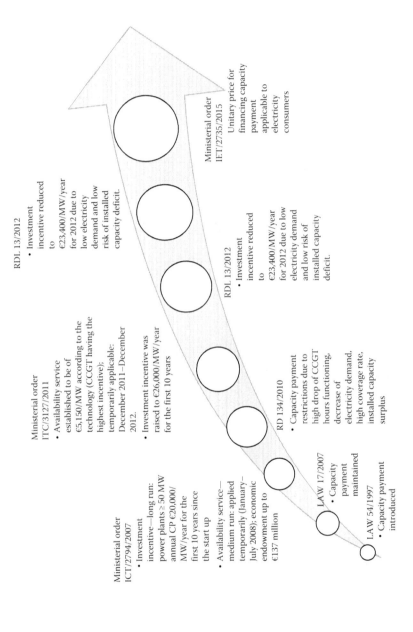

FIGURE 11.8
Capacity payment regulation.

11.5 Conclusions

The EU has designed different objectives within its growth strategy (EUROPE 2020). Among these priorities, the use of renewable energy is one of the main topics of discussion in Europe and Spain. Within the energy field, promoting renewable energy sources is, for Europeans and Spaniards, a must.

However, despite the common objective of increasing energy from renewable sources, many differences can be observed between each Member State's legal framework, as well as in the mechanism used for supporting renewables. In this sense, Spain is one of the leaders at the European level in promoting renewables, especially when it comes to wind and solar energy and its contribution to the electricity mix. This was possible due to a very strong support of these renewable energy sources by the national government.

Three different stages were detected in the Spanish support process of renewable energy sources: (1) a rather tentative support in the 1980s and the beginning of the 1990s in the initial stage, (2) the development stage since the second half of the 1990s until 2007, and (3) the slowdown stage from 2008 until today. It is the second stage that helped strongly promote renewable energy sources in Spain. Nevertheless, it is this legislation that, as a counterpart, contributed significantly to the creation of the tariff deficit, which led to an unsustainable situation that required a break from this development. Additionally, due to the discontinuity of renewable sources, security of supply was and still is one of the main concerns of all governments promoting renewables. In this case, Spain has to deal with disputes with traditional energy companies, mainly CCGTs. This is a consequence of the ambitious support for renewables against traditional sources as CCGTs have faced a huge limitation of their functioning hours. The legislation had to be designed bearing in mind the need to increase the share of renewable energy sources and their negative externalities (e.g., tariff deficit) and the security of supply, at the same time finding a proper balance between all these aspects. This has not yet been achieved by the different Spanish governments as they have faced grave issues in terms of tariff deficit, which has partially been overcome since 2014, when they started achieving surplus. Despite this timely surplus, tariff deficit is a problem yet to be solved. Unfortunately, the final consumers seem to be the ones paying for this mismatched strategy.

References

Agosti, L., Padilla, J., 2010. Promoción de las energías renovables: La experiencia de España. In: Moselle, B. et al. (Eds.), 2010: *Electricidad Verde*. Energías renovables y sistema eléctrico. Marcial Pons, pp. 517–542.

Azofra, D., Jiménez, E., Martínez, E., Blanco, J., Saenz-Díez, J.C., 2014. Wind power merit-order and feed-in-tariffs effect: A variability analysis of the Spanish electricity market. *Energy Conversion and Management* 83: 19–27.

Batlle, C., Pérez-Arriaga, I.J., 2007. Design criteria for implementing a capacity mechanism in deregulated electricity markets. IIT Working Paper IIT-07-024I. Available at http://citeseerx.ist.psu.edu/viewdoc/download?doi=10.1.1.473.242 5&rep=rep1&type=pdf. Accessed January 6, 2016.

Batlle, C., Vázquez, C., Rivier, M., Pérez-Arriaga, I.J., 2007. Enhancing power supply adequacy in Spain: Migrating from capacity payments to reliability options. *Energy Policy* 35(9): 4545–4554.

Bechberger, M., 2006. Why renewables are not enough: Spain's discrepancy between renewables growth and energy (in)efficiency. In: Mez, L. (Ed.), *Green Power Markets: Case Studies and Perspectives*. Multi-Science Publishing, Brentwood, U.K.

Buchan, D., Keay, M., Robinson, D., 2014. Energy and climate targets for 2030: Europe takes its foot off the pedal. The Oxford Institute for Energy Studies, London, U.K. Oxford Energy Comment. October 2014.

Burgos-Payán, M., Roldán-Fernández, J.M., Trigo-García, A.L., Bermúdez-Ríos, J.M., Riquelme-Santos, J.M., 2013. Costs and benefits of the renewable production of electricity in Spain. *Energy Policy* 56(May): 259–270.

Burrett, R., Clini, C., Dixon, R., Eckhart, M., El-Ashry, M., Gupta, D., Houssin, D., 2009. Renewable Energy Policy Network for the 21st Century. Renewable Energy Policy Network for the 21st Century. Renewables global status report; 2005–2010. http://www.ren21.net/globalstatusreport/g2010.asp. Accessed on July 1, 2016.

Bustos, M., 2004. The new payment mechanism of RES-E in Spain. Introductory report. APPA, Barcelona, Spain.

Caldés, N., Varela, M., Santamaría, M., Sáez, R., 2009. Economic impact of solar thermal electricity deployment in Spain. *Energy Policy* 37: 1628–1636.

Castro-Rodríguez, F., Marín, P.L., Siotis, G., 2009. Capacity choices in liberalised electricity markets. *Energy Policy* 37: 2574–2581.

Collado, E., 2012. Tecnologías: Aportación de la fotovoltaica. En Jornadas sobre generación distribuida y Balance Neto. Madrid, Spain, May 23, 2012.

Comisión Nacional de los Mercados y la Competencia (CNMC), 2014. Spanish Energy Regulator's National Report to the European Commission 2014. https://www.cnmc.es/Portals/0/Ficheros/Energia/Publicaciones_Anuales/140731_NR_CNMC_final.pdf. Accessed May 25, 2016.

del Río González, P., 2008. Ten years of renewable electricity policies in Spain: An analysis of succesive feed-in tariff reforms. *Energy Policy* 36: 2917–2929.

del Río González, P., 2010. Analysing the interactions between renewable energy promotion and energy efficiency support schemes: The impact of different instruments and design elements. *Energy Policy* 38: 4978–4989.

del Río González, P., Gual, M.A., 2007. An integrated assessment of the feed-in tariff system in Spain. *Energy Policy* 35: 994–1012.

Dinica, V., 2003. Sustained diffusion of renewable energy—Politically defined investment contexts for the diffusion of renewable electricity technologies in Spain, the Netherlands and United Kingdom. PhD thesis, Twente University Press, Enschede, the Netherlands.

Energy General Secretariat, 2001. Electricity and Gas Infrastructure Plan for 2002-2011. Reviewed version available at: http://www.minetur.gob.es/energia/planificacion/Planificacionelectricidadygas/desarrollo2002-2011/Documents/PLANeNERGETICA.pdf. Accessed July 15, 2016.

Energy Information Administration (EIA), 2007. Federal financial interventions and subsidies in energy markets. https://www.eia.gov/analysis/requests/2008/subsidy2/pdf/subsidy08.pdf. Accessed June 1, 2016.

ENTSO-E, 2015. European Network of Transmission System Operators for Electricity. https://www.entsoe.eu/Pages/default.aspx.

European Commission, 1996. Directive 96/92/EC of the European Parliament and of the Council of 19 December 1996 concerning common rules for the internal market in electricity, Brussels, Belgium. Accessed July 1, 2016.

European Commission, 1997. The energy for the future: Renewable energy sources. White paper for a Community Strategy and Action Plan. http://europa.eu/documents/comm/white_papers/pdf/com97_599_en.pdf. Accessed June 30, 2016.

European Commission, 2001. Directive 2001/77/EC of the European Parliament and of the Council of 27 September 2001 on the promotion of electricity produced from renewable energy sources in the internal electricity market. *Official Journal of the European Union* L283: 33–40. http://eur-lex.europa.eu/legal-content/EN/TXT/PDF/?uri=CELEX:32001L0077&from=EN. Accessed July 1, 2016.

European Commission, 2003. Directive 2003/54/EC of the European Parliament and of the Council of 26 June 2003 concerning common rules for the internal market in electricity and repealing Directive 96/92/EC2003/54/EC, Brussels, Belgium. *Official Journal of the European Union* L176: 37–56.

European Commission, 2005. The support of electricity from renewable energy sources. Communication from the Commission, Brussels, Belgium, 7.12.2005 COM(2005) 627 final. http://eur-lex.europa.eu/legal-content/EN/TXT/PDF/?uri=CELEX:52005DC0627&from=ES. Accessed July 1, 2016.

European Environmental Agency, 2004. Energy subsidies in the European union: A brief overview. Technical report 1. http://reports.eea.europa.eu/technical_report_2004_1/en/Energy_FINAL_web.pdf. Accessed July 1, 2016.

European Union, 2009. Directive 2009/28/EC of the European Parliament and of the Council of 23 April 2009 on the promotion of the use of energy from renewable sources and amending and subsequently repealing Directives 2001/77/EC and 2003/30/EC. *Off. J. Eur. Union* L140:6–62. http://eur-lex.europa.eu/legal-content/EN/TXT/PDF/?uri=CELEX:32009L0028&from=EN. Accessed on July 1, 2016.

Fabra, P.N., Fabra, U.J., 2012. El Déficit Tarifario en el Sector Eléctrico Español. *Papeles de Economía Española* 134: 88 100.

Folgado-Blanco, J., 2003. La política energética en España. *Información Comercial Española, ICE* 811: 29–33.

García, J.L., Menéndez, E., 2006. Spanish renewable energy: Successes and untapped potential. In: Mallon, K. (Ed.), *Renewable Energy Policy and Politics: A Guide for Decision Making*. Earthscan, London, U.K., pp. 215–227.

Gelabert, L., Labandeira, X., Linares, P., 2011. An ex-post analysis of the effect of renewables and cogeneration on Spanish electricity prices. *Energy Economics* 33: S59–S65.

Gil, H., Gomez-Quiles, C., Riquelme, J., 2012. Large-scale wind power integration and wholesale electricity trading benefits: Estimation via an ex post approach. *Energy Policy* 41, 849–859.

Haas, R., Panzer, C., Resch, G., Ragwitz, M., Reece, G., Held, A., 2011. A historical review of promotion strategies for electricity from renewable energy sources in EU countries. *Renewable and Sustainable Energy Reviews* 15(2), 1003–1034.

Held, A., Ragwitz, M., Haas, R., 2006. On the success of policy strategies for the promotion of electricity from renewable energy sources in the EU. *Energy and Environment* 17(6): 849–868.

IDAE, 1999. Plan for promoting renewable energy sources. Available at: http://www.idae.es/index.php/mod.documentos/mem.descarga?file=/documentos_4044_PFER2000-10_1999_1cd4b316.pdf. Accessed June 15, 2016.

IDAE, 2005. Plan de Energías Renovables 2005-2010 (Renewable Energy Plan 2005-2010). Available at http://www.idae.es/uploads/documentos/documentos_PER_2005-2010_8_de_gosto-2005_Completo.(modificacionpag_63)_Copia_2_301254a0.pdf. Accessed June 15, 2016.

International Energy Agency (IEA), 2010. Global renewable energy policies and measures database. http://www.iea.org/textbase/pm/grindex.aspx. Accessed July 10, 2016.

International Energey Agency (IEA), 2014. World Energy Outlook 2014. http://www.iea.org/publications/freepublications/publication/WEO2014.pdf. Accessed on October 20, 2016.

IPCC, 2007. Intergovernamental Panel on Climate Change. Fourth assessment report, Synthesis report. Accessed http://www.ipcc.ch. Accessed June 15, 2016.

Jacobsson, S., Bergek, A., Finon, D., Lauber, V., Mitchell, C., Toke, D., Vergruggen, A., 2009. EU renewable energy support policy: Faith or facts?. *Energy Policy* 37: 2143–2146.

Jensen, S., Skytte, K., 2003. Simultaneous attainment of energy goals by means of green certificates and emission permits. *Energy Policy* 31: 63–71.

Jones, C., 2010. Políticas de la UE para el desarrollo de las energías reno-vables. In: Moselle, B. et al. (Eds.) *2010: Electricidad Verde*. Energías renovables y sistema eléctrico. Marcial Pons.

Law 82/1980 (30th December) on Energy Conservation, https://www.boe.es/boe/dias/1981/01/27/pdfs/A01863-01866.pdf. Accessed June 5, 2016.

Law 54/1997 (27th November) of Electric Sector. https://www.boe.es/boe/dias/1997/11/28/pdfs/A35097-35126.pdf. Accessed June 5, 2016.

Law 34/1998 (7th October) on Hydrocarbons Sector. https://www.boe.es/boe/dias/1998/10/08/pdfs/A33517-33549.pdf. Accessed June 5, 2016.

Law 17/2007 (4th July) for modifying the Law 54/1997 (27th November) on the Electricity Sector in order to adapt it to the Directive 2003/54/CE of the European Parliament and European Council (26th June) on the common rules for the internal electricity market. https://www.boe.es/boe/dias/2007/07/05/pdfs/A29047-29067.pdf. Accessed July 1, 2016.

Law 24/2013 (26th December) of Power Sector. https://www.boe.es/boe/dias/2013/12/27/pdfs/BOE-A-2013-13645.pdf. Accessed July 7, 2016.

Liao, C.-H., Ou, H.-H., Lo, S.-L., Chiueh, P.-T., Yu, Y.-H., 2011. A challenging approach for renewable energy market development, *Renewable and Sustainable Energy Reviews* 15: 787–793.

Marín-Quemada, J.M. 2008. Política energética en la UE: el debate entre la timidez y el atrevimiento. *ICE Economía de la Energía* 842: 65–76.

Menanteau, P., Finon, D., Lamy, M.L., 2003. Prices versus quantities: Choosing policies for promoting the development of renewable energy. *Energy Policy* 31(8): 799–812.

Meyer, N.I., 2007. Learning form wind energy policy in the EU: Lessons from Denmark, Sweden and Spain. *European Environment* 17: 347–362.

Ministerial Order IET/2735/2015 (17th December) set access tariffs of electricity for 2016 and certain facilities and retributive type parameters production facilities are approved electricity from renewable energy sources, cogeneration and waste. Available at: https://www.boe.es/boe/dias/2015/12/18/pdfs/BOE-A-2015-13782.pdf. Accessed June 9, 2016.

Ministerial Order ITC/2794/2007 (27th September) reviews electric tariff from 1st October 2007. Available at: https://www.boe.es/boe/dias/2007/09/29/pdfs/A39690-39698.pdf. Accessed June 9, 2016.

Ministerial Order ITC/3127/2011 (17th November) regulates the service of power availability capacity payments and the incentive is modified investment referred to in Annex III of Order ITC/2794/2007 of 27 September, the electricity tariffs are revised from 1 October 2007. Available at: http://www.omie.es/files/orden_itc_3127-2011_de_17_de_noviembre.pdf. Accessed June 9, 2016.

Moreno, B., López, A.J., García-Álvarez, M.T., 2012. The electricity prices in the European Union. The role of renewable energies and regulatory electric market reforms. *Energy* 48: 307–313.

Muñoz, M., Oschmann, V., Tábara, D., 2007. Harmonisation of renewable electricity feed-in laws in the European Union. *Energy Policy* 35(5): 3104–3114.

Pimentel, D., Rodrigues, G., Wane, T., Abrams, R., Goldberg, K., Staecker, H., 1994. Renewable energy—Economic and environmental issues. *Bioscience* 44: 536–547.

REE, 2015. El Sistema Eléctrico Español. Available at: http://www.ree.es/sites/default/files/downloadable/avance_informe_sistema_electrico_2015_v2.pdf. Accessed June 17, 2016.

REN21, 2015. Renewables 2015 Global Status Report. Ed. Renewable Energy Policy Net work for the 21st Century. France. ISBN: 978-3-9815934-6-4 http://www.ren21.net/sta tus-of-renewables/global-status-report/.

Robinson, D., 2015. Análisis comparativo de los precios de la electricidad en la Unión Europea y en Estados Unidos: Una perspectiva española. Eurocofin. http://cdn.20m.es/adj/2015/10/20/3382.pdf. Accessed July 29, 2016.

Royal Decree 2366/1994 (9th December) generation of electricity through hydraulic installations, cogeneration and other from renewable energy sources. Available at: https://www.boe.es/boe/dias/1994/12/31/pdfs/A39595-39603.pdf. Accessed June 4, 2016.

Royal Decree 2818/1998 (23rd December) on Special Regime of generation of electricity for installations supplied by renewable energy sources, waste and cogeneration. Available at: https://www.boe.es/boe/dias/1998/12/30/pdfs/A44077-44089.pdf. Accessed June 4, 2016.

Royal Decree 436/2004 (12th March), which establishes the methodology for the update and systematization of the legal and economic regime of electricity production under special regime. Available at: https://www.boe.es/boe/dias/2004/03/27/pdfs/A13217-13238.pdf. Accessed June 9, 2016.

Royal Decree 661/2007 (25th May), which regulates the electricity production under special regime. Available at: https://www.boe.es/boe/dias/2007/05/26/pdfs/A22846-22886.pdf. Accessed May 3, 2016.

Royal Decree 1578/2008 (26th September) on remuneration for electricity production from solar photovoltaic technology for technologies installed after the maintenance of the remuneration of the Royal Decree 661/2007. Available at: https://www.boe.es/boe/dias/2008/09/27/pdfs/A39117-39125.pdf. Accessed January 8, 2016.

Royal Decree 134/2010 (12th February) on establishing the resolution procedure to guarantee supply restrictions and amendment of the Royal Decree 2019/1997, of 26 December, which organizes and regulates the market for electricity production. Available at: https://www.boe.es/boe/dias/2010/02/27/pdfs/BOE-A-2010-3158.pdf. Accessed January 9, 2016.

Royal Decree 900/2015 (9th October), which regulate administrative, technical and economic conditions of the supply modalities of electricity from self-consumption and of the production with self-consumption. Available at: https://www.boe.es/boe/dias/2015/10/10/pdfs/BOE-A-2015-10927.pdf. Accessed February 1, 2016.

Royal Decree 947/2015 (16th October), which establishes a call for granting specific remuneration system for new installations producing electricity from biomass in the mainland electricity system and for wind technology. Available at: https://www.boe.es/boe/dias/2015/10/17/pdfs/BOE-A-2015-11200.pdf. Accessed February 2, 2016.

Royal Decree Law 6/2009 (30th April) on Energy Sector. Available at: https://www.boe.es/boe/dias/2009/05/07/pdfs/BOE-A-2009-7581.pdf. Accessed March 9, 2016.

Royal Decree Law 6/2010 (9th April) on measures for promoting economic recovery and employment. https://www.boe.es/boe/dias/2010/04/13/pdfs/BOE-A-2010-5879.pdf. Accessed March 9, 2016.

Royal Decree Law 1/2012 (27th January) on suspension of procedures for pre-allocation and the abolition of economic incentives for new installations producing electricity from cogeneration, renewable and waste energy. Available at: https://www.boe.es/boe/dias/2012/01/28/pdfs/BOE-A-2012-1310.pdf. Accessed June 9, 2016.

Royal Decree Law 13/2012 (30th March) on transposing directives on the internal markets in electricity and gas and electronic communications, and measures for correcting deviations mismatch between costs and revenues of the electricity and gas sectors. Available at: https://www.boe.es/boe/dias/2012/03/31/pdfs/BOE-A-2012-4442.pdf. Accessed June 9, 2016.

Royal Decree Law 20/2012 (13th July) on measures to ensure fiscal stability and promoting competitiveness. Available at: https://www.boe.es/boe/dias/2012/07/14/pdfs/BOE-A-2012-9364.pdf. Accessed June 9, 2016.

Royal Decree Law 9/2013 (12th July) on urgent action to ensure financial stability of the electrical system measures. Available at: https://www.boe.es/boe/dias/2013/07/13/pdfs/BOE-A-2013-7705.pdf. Accessed June 9, 2016.

Ruiz Romero, S., Colmenar Santos, A., Alonso Castro Gil, M., 2012. EU plans for renewable energy. An application to the Spanish case. *Renewable Energy* 43: 322–330.

Sáenz de Miera, G., del Río González, P., Vizcaíno, I., 2008. Analysing the impact of renewable electricity support schemes on power prices: The case of wind electricity in Spain. *Energy Policy* 36: 3345–3359.

Sallé-Alonso, C., 2012. El déficit de tarifa y la importancia de la ortodoxia en la regulación del sector eléctrico. *Papeles de Economía Española* 134: 101–116.

Sebastián, M., 2015. La falsa bonanza. Cómo hemos llegado hasta aquí y cómo intentar que no se repita. Grupo Editorial 62, Barcelona, Spain.

Sevilla, J.M., Golf, E., Driha, O.M., 2013. Las energías renovables en España. *Estudios de Economía Aplicada* 31(1): 35–58.

Solorio, I., 2011. La europeización de la política energética en España: ¿qué sendero para las renovables? *Revista Española de Ciecnias Políticas* 26: 105–123.

Swider, D.J., Beurskens, L., Davidson, S., Twidell, J., Pyrko, J., 2008. Conditions and costs for renewables electricity grid connection: Examples in Europe. *Renewable Energy* 33: 1832–1842.

Index